Water Governance – Challenges in Africa

Hydro-optimism or hydro-pessimism?

CEAUP Studies on Africa

Volume 2

Series editors:

Alexander Keese
João Carlos Garcia
Maciel Santos

Centro de Estudos Africanos U. P.
Faculdade de Letras da Universidade do Porto
Via Panorâmica s/n
4150-564 Porto

PETER LANG
Bern · Berlin · Bruxelles · Frankfurt am Main · New York · Oxford · Wien

Ibrahima Anne (ed.)

Water Governance – Challenges in Africa

Hydro-optimism or hydro-pessimism?

PETER LANG
Bern · Berlin · Bruxelles · Frankfurt am Main · New York · Oxford · Wien

Bibliographic information published by die Deutsche Nationalbibliothek
Die Deutsche Nationalbibliothek lists this publication in the Deutsche Nationalbibliografie; detailed bibliographic data is available on the Internet at ‹http://dnb.d-nb.de›.

British Library Cataloguing-in-Publication Data:
A catalogue record for this book is available from The British Library, Great Britain

Library of Congress Cataloging-in-Publication Data

Water governance–challenges in Africa : hydro-optimism or hydro-pessimism? / Ibrahima Anne (ed.).
p. cm. – (CEAUP studies on Africa ; v. 2)
Conference proceedings.
Includes bibliographical references.
ISBN 978-3-0343-1133-5
1. Water-supply–Africa–Management–Congresses. 2. Water-supply–Africa–Management–International cooperation–Congresses. 3. Water resources development–Africa–Congresses.
4. Water resources development–Africa–International cooperation–Congresses.
5. Watershed management–Africa–Congresses. 6. Water-supply–Government policy–Africa–Congresses. I. Anne, Ibrahima, II. Series: CEAUP studies on Africa ; v. 2.
HD1699.A1W385 2012
333.910096–dc23

2012043390

We would like to thank Professor Ramiro Pimenta for his valuable collaboration and his meticulous text revision.

Cover illustration: Western Sahara woman in the refugee camp of Tindouf, Algeria by Mohamed Massara. Agencia LUSA, Portugal.

ISBN 978-3-0343-1133-5 pb.
ISSN 1663-3059 pb.

ISBN 978-3-0351-0518-6 eBook
ISSN 2235-591X eBook

Printed in Switzerland

Alan Nicol[*]

Prologue: A starting point...

Africa's waters are 90% shared. At its core future water management and development is a collective challenge straddling national, river basin and socio-cultural boundaries. In many cases boundaries overlap, are contiguous or cross each other at right angles. Grounds for optimism or pessimism over the future for 'water in Africa' are very much embedded in this complex decision making space, sometimes referred to as the 'problemshed'.

This conference enabled a range of researchers and practitioners to convene and explore key social, cultural and political dynamics around both large and small water development in Africa. This distinction helps in unpacking the debates that surround water for food production, hydropower development, and the smaller in quantity but no less important water that households need for future health and productivity.

The eleven papers in this conference covered much of this ground and, therefore, provide an especially exciting intellectual contribution to current understanding. They shed light not only on the scale of the collective challenge, but also on the qualitative fault lines that lie within from role of the market economy to the meaning and interpretation in policy and practice of the human right to water.

Analyses of the complexity of water governance including intricate social-ecological systems and processes of decentralisation are explored, as are the opportunities afforded by focusing on new forms of learning. We are brought up to date with realities of new external actors influencing Africa's water management through a focus on Chinese assistance to dam building, and then brought firmly back to ground-level and local realities with an exploration of the 'small water' issues involved in water and poverty linkages. We learn of experiences in integrated water management and

* London. Knowledge, Technology and Society Team. Institute of Development Studies. University of Sussex, Brighton. United Kingdom.

of hydropolitical 'risk' as well as the potential contribution of rainwater harvesting to Africa's future agricultural growth – a pillar for future development across the continent. Overall the richness and depth these papers provide should be a stimulus to further policy debate and critical thinking on future development scenarios.

And this is a task just in its infancy. Already to the mix we need to add climate change scenarios and the impact of continued high population growth. The research agenda is growing, and, rightly so, increasingly being led from Africa and by African institutions. These papers can aid in a process of converging research efforts both North-South and South-South, which will bridge the optimism-pessimism divide and seek to support better planning and investment decision making that takes a more informed approach and one build on concepts and practice of no or low-regret decision making. Whilst the hydro-optimists might say that the right decisions are being taken to 'futureproof' Africa against natural and man-made hazards including droughts and floods (and deterioration in water quality), for the pessimists the future is always frightening. For researchers of either bent, the future remains an open question, and our job is to continue the search for answers.

Contents

List of Maps

Ibrahima Anne*

Introductory notes

The book represents the outcome of a conference organised by Centre of African Studies of the University of Porto in 2008, which aimed at understanding the current situation and debating the challenges related to the management of water resources in the African continent. Under the title 'Water in Africa: Hydro-Pessimism or Hydro-Optimism?', the conference has gathered several international experts to debate several water-related topics. Being water a multifaceted resource – both a natural, economic, social and political resource – the issue was approached by several perspectives.

The common denominator between the experts presenting and attending the conference was the attempt to answer some of the central and critical questions:

- How are water resources being managed in Africa?
- What are the main challenges concerning access, distribution and allocation of freshwater resources in the continent?
- What are the driving forces that shape current and future demand for water?
- What are the solutions being developed by communities, policy-makers, private sector and environmental activists?
- What are water scientists and international donor organizations contributing?

In answering these questions, the experts were also required to think along the optimism and pessimist lines, and assume a position about the current and future water situation in the African continent. Pessimists highlight

* Centro de Estudos Africanos da Universidade do Porto (CEAUP); Degrémont SA, Rua General Ferreira Martins nº8, 3ºA/C/D, Miraflores, 1495–137 Algés, Portugal. anne.ibrahima@gmail.com.

water scarcity and insecurity, water inequity and the conflict potential at local and regional levels. They forecast a gloomy future for Africa. Can they be wrong? Optimists emphasize the potential of improved water management practices, water policy reforms and socially, economically and environmentally sustainable policies. They argue that local adaptive capacities can be enhanced and regional hydropolitical cooperation can be established. Water optimists believe in windows of opportunity to improve water resources management and that Africa's water crises can be addressed. Can they be right?

The conference had been organized around four major topics that dominate the water sector in Africa as well as the research and on the field. These same topics constitute the four sections of this book.

- Water as a Human Right
- Water Governance
- Water Supply and Sanitation
- Hydropolitical risks and Transboundary water resources policy

In an original and rather direct way, the book presents a broad view of water-related problems in the African continent. These problems are not so Africa-specific, but they can also be considered as typical for the continent.

In one hand, it is known that all continents to some extent suffer from water related problems, some more than others; on the other hand, the economic context of the African countries is more difficult, compared to all others. In view of the demographic structure, food, water supply and sanitation shortage in the continent, water aspect may well represent the most limiting factor for the development of the African continent.

Water governance

The challenges of developing Africa's water resources are daunting. For instance, it is estimated that nearly 40% of Africans do not have access to safe water and adequate sanitation facilities. Besides, the level of water resource exploitation is low with only between 3%–5% of the total amount used under managed conditions. Worse still, per capita water storage capacity in Africa is 100 times lower than in Europe and North America (6150 m3/person compared to Ethiopia at 47 m3/person), leading to lack of water security for social and economic development and greater vulnerability to water related disasters. Africa is home to more than 60 shared river basins but weak regional cooperation has limited the benefits to the continent and reduced the effectiveness of water governance necessary for development. These challenges are compounded by steady environmental degradation, depletion and contamination of water resources in addition to related factors such as climate change, desertification, flooding and erosion.

It is estimated that that an average investment of US$ 20 billion is required each year over the coming 25 years to attain the water Millennium Development Goals (MDG) and African Water Vision targets (AfDB).

The African Water Vision envisages "An Africa where there is an equitable and sustainable use and management of water resources for poverty alleviation, socioeconomic development, regional cooperation and the environment" while the MDG for water seeks to "Reduce by half the proportion of people without sustainable access to safe drinking water" in the world by 2015.

The Rural Water Supply and Sanitation Initiative (RWSSI) is one of the main initiatives of the African Development Bank to help achieve the MDG in Africa. It was initiated in 2002 to address the problem of low access to water supply and sanitation in rural Africa where the majority of the population lives.

It must be fully acknowledged that "Meeting the Millennium Development Goals on Water and Sanitation" is one of the biggest *challenges faced by African countries.*

The African Water Facility (AWF) is an initiative led by the African Ministers' Council on Water (AMCOW) to mobilize financial resources for water sector development in Africa. Created in 2004, the AWF is hosted and managed by the African Development Bank (AfDB).
Major objectives of the African Water Facility include attracting and making effective use of increased and appropriate investments needed to achieve national and regional water sector targets in Africa.

Human right to water

Despite these efforts, there is still ongoing an international debate on the necessity of the existence of human right to water. However, there is a great deal of doubt and uncertainty on whether that recognition will have practical implications, especially in African continent.

There are also legitimate doubts about the ability of the market in providing everyone access to safe water in Africa, and some may see a conflict between the market approach and the concept of right to water in the continent.

However, the low level in institutional development and lack of political will and governance transparency may be at the origin of state incompetency in regulating the market and guaranteeing full coverage of safe water. In this case, there exists a necessity of specific policies to guarantee minimum standards.

Thus, there is a great interest in assessing the politics of water management in relation to the institutional context leading to water privatization in Africa. This process is said to be resulting from a failure of public water supply by the state.

Water supply and sanitation

In fact, it is a disturbing situation that many African countries face in front of the attainment of the MDG on water and sanitation. More than a million Africans die every year due to lack of hygiene, sanitation and access to safe drinking water, whereas the international community had committed, in September 2000, to reduce by half by 2015, the number of Africans without access to safe drinking water. This situation has been compounded by environmental degradation and the adverse effects of climate change.

In some countries, deficiency in access to water is so severe that it allows for reappearance of water-borne diseases, associated mainly with poverty. These are clearly examples of mismanagement and are present in many countries, especially the poorest and/or coming out of recent periods of political unrest or armed conflicts.

There must be an awareness of the fact that access to water and control over its management in arid zones generates necessarily local political struggles. The study of conflicts over access and control of water resources, like for any other resource, can help understand practices and processes of local governance.

Demographics and human pressure on water resources in arid and semi-arid zone, that represent a huge proportion of lands in Africa, are pushing these resources to their limits. Often, the solution of infrastructural development and institutional reforms can create fundamental change of water management from a mainly customary form of organisation to a hybrid system of institutions: state, market and community-based stakeholders. The most interesting is to evaluate and investigate how the social-ecological system will be able to absorb these external changes in term of its resilience. Hence, water-sector reforms via new legislation must be taken to increase equity, efficiency and environmental sustainability of natural water resource use, in view of an integrated water resource management.

Transboundary water resources

Worldwide, most of the surface water and groundwater resources are shared by two or more countries. The African continent is not an exception. Most of Africa's river and lake basins crisscross administrative borders of several countries, and as such are considered transboundary watercourses. According to the Oregon Water Database, within the already referred large number of transboundary river basins in Africa, some are shared by plenty of riparian countries – the Nile by 10, the Niger by 9, the Zambezi by 8. Some others are shared by less countries, but hydrology, geography, history and/or climate variability can make the management of the water resources equally complex. The challenge of managing transboundary water bodies is immense as it involves complex political and legal processes, usually marked by complex hydropolitics and power relations between the riparians. Every basin is unique – thus the management and allocation of its water resources is also unique. The last section of this book addresses transboundary water resources in major African river basins.

From problems of access and control over water resources, to governance and sound integrated water resource management, the path seems long and difficult.

PART I
Water as a human right

The market economy and the human right to water in Africa

Implementation of the human right to water: multi-level challenges

Manuel Couret Branco* and Pedro Damião Henriques**

The market economy and the human right to water in Africa

Abstract. This paper addresses concerns the failure of the market to provide universal access to clean water in Africa. More specifically, this paper examines the conflicts that may occur between the market discourse and the right to water on this continent.

First of all, market hegemony hinders the human right to water due to market inefficiencies which interfere with universal water provision. On the other hand, it is an unaccountable system and human rights purveyors need by principle to be submitted to democratic control.

Secondly, lower levels of socio-economic development tend to correlate with less governmental ability to supply basic goods, regulate markets and guarantee universal access to safe water. This is the situation of many African countries, which requires specific policies in order to guarantee a minimum satisfaction of basic human rights, in particular water supply.
Keywords. Human Rights, Africa, Right to Water, Economic Theory, Market.

Introduction

Life begins with water. No organism can live without it. As a matter of fact, all life forms, humans included, live in an aqueous environment. But

* Universidade de Évora, Departamento de Economia, NICPRI, Largo dos Colegiais, 2, 7000 Évora – Portugal; mbranco@uevora.pt.

** Universidade de Évora, Departamento de Economia, CEFAGE, Largo dos Colegiais, 2, 7000 Évora – Portugal; pdamiao@uevora.pt.

water also happens to be an economic, social and cultural good that is used to satisfy the ever-growing range of human needs. Indeed, since 1950, as the world population has doubled, water use has tripled. Water scarcity, in both its quantitative and qualitative manifestations is, therefore, emerging as a major development challenge for many countries. Furthermore, the lack of water hampers development by constraining food production, health conditions and industrial development (UNECA, 2006).

About 2% of the earth's blue surface visible is fresh water, the rest being salt water. Furthermore, only about half of the fresh water is available for use by human beings. Africa's share of global freshwater resources is merely about 9% of the world total and this limited amount is distributed unevenly across the continent, with Western and Central Africa having significantly greater precipitation levels than the Northern and Southern regions of the continent. The Democratic Republic of Congo, for example, has 25% of African fresh water resources whilst Mauritania retains only 0.001%. The lack of rainfall means that many countries rely on groundwater: 60% of Algerian needs are met in this fashion, 95% of Libyan requirements and 15% for the continent as a whole (2003 IYFW, 2008).

Despite regional variations in water resources, safe water availability across Africa falls far short of international standards. For instance, between 1990 and 2004, access to improved water in Sub-Saharan Africa increased merely by 7%, from 48 to 55%. The situation is much worse for the population using improved sanitation, which increased from 32 to 37 per cent. In the same period, the African population living below the poverty line decreased by 5%, an improvement that has still kept, however, 40% of the population living below that line (UNDP, 2006).

As a consequence, in many African countries water users have not being paying the true costs of supplying water and it is unlikely that they will have the conditions to meet those costs in the foreseeable future. Indeed, only a small proportion of the cost of the transport, treatment and disposal of water has been supported by users, the remaining having been subsidized. Governments have, so far, proven obviously incapable of meeting the basic water needs of most of the population, whether in rural or in urban areas.

In order to improve the livelihood of African populations, water supply is on the front line of the services that should be provided. Credit and donor institutions maintain that it can be delivered either by governmental or market mechanisms. This paper analyses the potential role of markets in meeting African peoples' right to water.

The right to water

The Universal Declaration of Human Rights states in its Article 3 the inalienable right to life. Other articles extend the this right beyond that of mere survival, demanding that life should also include the guarantee of minimum standards of human dignity and that it should be enjoyed with freedom and safety. Indeed, this right to life demands unquestionable access to both indispensable natural resources and manufactured goods. Petrella (2004) has argued that such natural resources should be classified as the minimum basic capital for human existence and, thus, specific policies regarding their exploitation and distribution should be instituted. Water should probably be the first on the list of these basic resources

The European Declaration for a New Water Culture signed in Madrid in February 2005 argues, in turn, that water should be classified according to four main categories. The first three, *water-for-life*, *water-for-citizenship* and *water-for-business* (Arrojo, 2006: 104–105) are of interest here. Indeed, adequate supplies of safe water are necessary to prevent death by dehydration, to reduce the risk of diseases and to attend to many other sorts of indispensable needs like farming or manufacturing, cooking or personal and domestic hygiene, to which one should also add a wide range of cultural needs such as the performance of religious rites or the plain enjoyment of leisure. The first two categories in the classification proposed by the European Declaration for a new Water Culture could, then, be included in the broader category of water as a human right and the last in water as an economic good.

As a matter of fact, several claims have been issued in the last decade stating that water ought to be considered a human right. On April 2, 1998, for example, a group of international personalities, such as the former presidents of Portugal and Argentina, Mário Soares and Raúl Alfonsín, respectively, issued a manifesto declaring water to be a common good belonging to all inhabitants on earth and an inalienable individual and collective right. In response to that and to other pleas, and following upon the International Covenant on Economic, Social and Cultural Rights (ICESCR), the United Nations Committee on Economic, Social and Cultural Rights proclaimed in November 2002 the Right to Water as a substantive implication of the implementation of the ICESCR, resulting from an extensive interpretation of its Articles 11 and 12. For the record, the first of these two articles recognizes the right of all peoples to an adequate standard of living, and the second the right of everyone to the highest attainable standards of physical and mental health.

This document declares water to be a limited natural resource and a public good fundamental for life and health; it further argues that the human right to water is indispensable for leading a life of human dignity and that it is a prerequisite for the realization of other human rights (UN, 2002: 1). Further justification for classifying water as a human right appears in article 15 of the ICESCR recognising the right of all people to participate in cultural life. Many religious rites require the use of water. According to this committee, the human right to water entitles everyone to sufficient quantities of safe, acceptable, physically accessible and affordable water for personal and domestic use.

The right to water, therefore, as in the case of most economic, social, and cultural rights, has both a quantitative and a qualitative dimension. From the quantitative point of view, it is stated that everyone should be provided with an amount of water sufficient to meet human needs according to WHO parameters. This quantity is not easy to determine because it can vary according to cultural idiosyncrasies and geographical location, but it is assumed that a person needs a minimum of 20 litres per day (UNDP, 2006: 8). This amount seems ridiculous when citizens in western developed countries spend more than that in just flushing their toilets. However, this meagre quantity is still unavailable to many on the planet. From the qualitative point of view, it is, in turn, stressed that the water provided should be safe, in other words, its consumption should not put human health at risk. As in the case of all human rights, no discrimination based on gender, religion, or social condition, can be tolerated regarding access to said water.

Now, if water can be classified as a public or a common pool good, and tap water as a private good, the entire process of providing safe water to people displays a dual character. Once the right to water is accepted as a basic human right, one is forced to admit its public character. Indeed, if water constitutes a human right because it is essential to life and a prerequisite for the enjoyment of other human rights, the excludable character of private goods means, therefore, that it is possible for an individual to be deprived of a human right on the basis of purchase power. This immediately transforms the lack of access to water into a violation of citizens' rights and, consequently, into a major political issue.

The UN declaration on the right to water states that people should have the means to access water (UN, 2002: 6). This signifies that it is acceptable for water to have a price and therefore to be submitted to economic principles. However, can it be inferred that the market is qualified to guarantee the human right to water? Indeed, there are reasons to believe that the market alone is not theoretically equipped to meet this challenge without abdicating an important set of constitutive principles.

The right to water in Africa

Despite humans' essential need for water, access to it is far from universally guaranteed and, amongst those communities which benefit from this access, water is far from equitably distributed. The World Health Organization believes that more than a billion people are deprived of basic access to water (WHO, 2001: 1). Three hundred million of the estimated 800 million African inhabitants live in a water-scarce environment, mostly in the northern and southern regions. On average, 64% of the African population has access to an improved water supply. However, about two thirds of the African population lives in rural areas where water coverage is the poorest, supplying only about 50% of the population. In urban areas an estimated 86% of the population has access to safe water.

However, people in these areas face two main problems. Many urban centres have water distribution systems in decline due to inadequate, aging and overloaded networks; and peri-urban dwellers live in slums and are impoverished, many of whom lack reasonable access to safe water (2003 IYFW, 2008).

The United Nations Organization estimates that approximately 2.3 billion people suffer from diseases associated with water, either its shortage and/or poor quality (UN, 1997: 39). In Africa, around 3 million people die annually as result of water related diseases (UNEP, 2002). Almost half of Africa's population suffers from one of six major water related diseases. Every day, 650 people die from diarrhoea, for instance, mainly children under five years of age. Many of theses diseases, such as schistosomiasis or cholera, seem to have become an African specialty, 82.8% and 78% respectively of the world cases occurring in Africa. Malaria, for example, is the leading cause of disease in young children and constitutes 10% of the overall disease burden, slowing African economic growth by 1.3% per year (2003 IYFW, 2008; WWF, 2002).

It is argued herein that classifying the right to water as a basic human right should lead to improvements in water resource management and drastically reduce the number of people without access to safe water and adequate sanitation. The resultant expansion of irrigation and enhanced control of droughts, floods and desertification should bring further economic development. There are ample resources available which, if developed and managed in a sustainable way, will enable Africa to reach the goals set within the framework of the Millennium Development Goals (MDGs) and the Africa Water Vision 2025. This calls for an increase in the water resource potential of 10% in 2015 and by 25% in 2025 and a reduction of 50% in the number of inhabitants without access to safe water (UNDP, 2006).

Expanding water supply is not only a question of guaranteeing a basic human right; it is also a development question. Indeed, improvements in water supply and sanitation in Africa will contribute decisively to achieving the MDGs, improving the livelihood of African people by eradicating extreme poverty and hunger, achieving universal primary education, promoting gender equality and empowering women, improving health conditions (reducing child mortality, improving maternal health and combating HIV and water diseases), ensuring environmental sustainability and developing a global partnership for development. Water provision can be seen as a prerequisite for the realisation of the human right to education for example. Indeed, women and children travel 10 to 15 km, on average, spending eight or more hours per day collecting water and carrying up to 15 litres per trip (Conteh, 2006). If one considers the opportunities lost to water collection, insufficient water coverage decisively contributes to poverty upon adulthood. As a result, the gender development index is 0.513, meaning that Africa's female population is 0.020 less developed than the male population.

Achieving the MDG will only be feasible if African governments implement policies that make access to water a human right, increase investment in water facilities for rural and urban areas, regulate water supply in order to improve efficiency and enhance equity and introduce integrated water resource management to promote sustainability (UNDP, 2006). As far as the meeting of basic needs is concerned, one can safely state that it is relatively easy and cheap to provide universal access to water. Pedro Arrojo, a leading water ethics scholar and recipient of the Goldman Environmental Prize in 2003, declares that providing water for people's basic needs is within reach of every national economy (Arrojo, 2006: 109). Furthermore, the 2006 Human Development Report states that the direct and indirect costs of the current safe water deficit in developing countries represent nine times that of providing universal coverage. The overall loss due to lack of water and sanitation is about 5% of GDP and each monetary unit spent produces 8 units in savings and productivity (UNDP, 2006).

Why, then, is there such inequality in water distribution? In many developing countries where access to water is guaranteed, inequality remains due to pricing, as described above. Indeed, poorer households often pay more for water than richer ones. According to the above-mentioned Human Development Report, households living in African slums pay five to ten times more for water than wealthier households in developing world cities (UNDP, 2006: 10). One of the reasons for this discrimination arises from the fact that many poor households do not have access to public systems of water distribution, owing to numerous factors, such as inability to pay and commercially biased infrastructural investments, for example.

Examples of public inability to achieve universal water coverage in Africa include Dar-es-Salam, Tanzania, or Ouagadougou, Burkina-Faso. In the latter case, less than 30% of the population is connected to the public water distribution system (UNDP, 2006: 9), frequently because water is still too expensive for poor households (UNDP, 2006: 10). Therefore, market supply of water has been presented as an alternative to public distribution in order to guarantee the right to water. Actually, since the 1980s, privatisation has been considered an African economic panacea in large part due to the conditions attached to IMF, IFC and WB loans and debt relief programs. Privatization was also a central component of donor-funded aid programs financed by development agencies of the developed countries. Water privatisation in Africa has taken different forms: contracts on a fee for service basis, management contracts, lease agreements, concessions, build-operate-transfer agreements and divestitures. The first three forms of privatisation are short-term and do not involve responsibility for capital investment (Conteh, 2006). In this process, French multinational companies have dominated privatisation (Bayliss and Hall, 2000).

The arguments in favour of water privatisation in Africa were to enhance operational efficiency, economic growth and development of the water sector, to respond to the inability of public utilities to raise capital investment and to deliver fiscal benefits. Detractors of privitisation argued that governments were abdicating their obligations to provide essential life services. There is little practical evidence that privatisation does, in fact, result in increased efficiency, economic growth and/or development. Private firms are interested in profit not in social objectives. In the final analysis, the end users are unable to pay the tariffs required by privatised companies. Regardless of this factual inability of privatisation to ensure the human right to water, we will try to show how the market alone is not theoretically able to ensure any human right, let alone the right to water.

Market hegemony versus the right to water

One of the crucial questions is to determine which type of institution is better qualified to ensure every citizen's right to water. State inefficiency in delivering water to all has, most especially in developing countries, constituted the main argument for asserting that the market should play a more active role in providing essential goods and services. As a result, we have witnessed extensive privatization of water supplies in developed as well as developing countries. The following pages will argue that the market is not fully equipped to guarantee the supply water as a human right. Firstly, the market does not voice social preferences; secondly, it lacks accountability; thirdly, it is inefficient; and finally, the water market can never be a competitive one.

The market does not voice social preferences

First of all, when universal rights such as human rights are being promoted, one is asserting a social preference. In the case of the right to water, one is therefore inclined to admit that the degree to which people's needs are being met in one set of circumstances is preferable vis-à-vis another. For instance, a situation in which all the population benefits from safe tap water is better than any other. Actually, once water is defined as a human right, universal coverage is the only acceptable target situation. Any other condition must therefore be considered not only inferior, but also unacceptable, as it would constitute a violation of people's human rights.

In this sense, the market would have a hard time promoting the human right to water in Africa or anywhere else simply because it is incapable of voicing social preferences, such as structural preferences concerning, for example, income distribution or water coverage. As a result of all the information conveyed by economic agents, the market can utter many preferences, as to, for instance, quantities of water to be produced, how and when, but it does not have arguments to assert that universal coverage is better than any other pattern of water distribution. What matters for the market is that agents are satisfied, in other words that sellers are able to sell the amounts they wish at market prices and that buyers are able to buy what they intend at the same market prices. The fact that some agents are not able to buy what they wish at market prices on account of excessively tightened budget constraints is of hardly any concern. In terms of private goods this may be generally acceptable, but when private goods are taken as rights, as in the case of tap water, exclusion becomes intolerable.

That is why, despite the fact that public distribution of water has been unable to meet the needs of the people, private supply does not appear to be a viable alternative. Indeed, if in Senegal for example, results of privatization can be described as mixed, meaning that household connections to water services have consistently risen although many of the poorer households remain unconnected because of costs (Conteh, 2006: 36), in South Africa privatization has proven to be a poor alternative to public distribution. One case in point occurred in Bhofolo, a black township in the Eastern Cape province, where water prices increased by 300% for township residents in 1995. In 1996 they increased again by 100% (Conteh, 2006: 37). These high tariffs, together with connection fees, meant that many households could not afford water, being either disconnected from water services or unable to get connected to the water supply system (Conteh, 2006: 38). The same general pattern occurred in many other townships such as Mlungusi, or Nelspruit.

Comparative historical studies of water supplies can partly explain why the market fails to efficiently promote the right to water in poor countries. Private companies supplying water in developed countries have inherited a heavy infrastructure financed by past public investments, supplying universal coverage to an average high-income market. In developing countries, on the other hand, limited and frequently damaged infrastructure, low levels of connection and high levels of poverty increase the tensions between business profitability and the supply of water at a fair price. In Buenos Aires, Argentina, for example, the water concession holder managed to expand the connections to the supply system, but at a slower pace than agreed in the concession contract because progress was slower in the poorer areas of the city. In Jakarta, Indonesia, likewise, three quarters of the new connections concerned medium and high income households or private and public institutions (UNDP, 2006). As a matter of fact, Pedro Arrojo (2006) found that multinational companies, holding most of the world's privatized concessions, may be interested in water distribution management but not in infrastructural investment, disregarding, therefore, one of the pre-conditions of the right to water which is universal access to said water.

The market is not accountable

When speaking of human rights, the rights of individuals correspond to duties of other individuals. In other words, human rights represent the rights which individuals have over the conduct of others. Therefore, if the rights of some individuals are not ensured, this is due to the fact that other individuals or institutions have failed to carry out their duties. When the state, for example, fails to ensure a citizen's human rights, the state is accountable either legally, in a court of law, or politically through elections. If the market fails to ensure human rights, whom should an individual turn to? The state is both elected and known, the market, on the other hand, is anonymous by definition. In this sense, the market is therefore not only unequipped to ensure the right to water in particular, but also to allocate rights in general. Furthermore, if the behaviour of agents or institutions represents an obstacle to the assertion of human rights, then it is their duty to change it. One can thus deduce that individuals have rights over the design of economic and social arrangements, namely over the definition of economic systems as far as they affect the assertion of human rights.

This conception of human rights reveals yet another contradiction with the language of economics. In economics, deprivation has been seen as the outcome of either nature's random behaviour or human incompetence. In other words, deprivation resulted either from nature playing nasty tricks on people or people being incapable of making the right decisions in addressing basic economic problems. The search for the good life signified, therefore, a struggle to dominate nature or to predict and mitigate its whims, and a quest for efficiency in human action. The rhetoric of human rights, in contrast, introduces a substantially different approach to deprivation by transforming economic problems into potential violations of citizens' rights, that is to say, into discrimination or structures that prevent people from exercising rights (Offenheiser and Holcombe, 2003: 275).

As a matter of fact, responsibility and democratic control of water suppliers has recently proved to be a key element in achieving universal water coverage. In Porto Alegre, Brazil, for instance, water services were private until 1904; then the city took them over. In 1961 it created a deliberative council which has represented many sectors in overseeing the DMAE (Municipal Department of Water and Sanitary Sewage) and played an important part in its success (Maltz, 2005: 34). The implementation in 1989 of the participatory budget, a municipal management system created specifically in Porto Alegre and within which the people gather throughout the year to decide where the investments are going to be made, brought DMAE even closer to the society and established a new level in control over the utility (Maltz, 2005: 34).

As one might expect, people are mainly interested in obtaining wider access to water and sanitation and, thus, between 1989 and 1996, the number of households with access to water rose from 94.7% to 99.5% while the percentage of population served by the municipal sewage system rose from 73% to 84% (Maltz, 2005); these figures are the highest in Brazil.

The market is inefficient

As we have already mentioned, water is used for a wide range of purposes, from human consumption to production activities such as transportation, industry, agriculture and fishing, as well as cultural, recreational, leisure, conservation and environmental activities. Taking into consideration the diversity of uses and the indispensability of water for satisfying basic human needs, a new question arises: how to prioritise the different types of water demand. A competitive market allocates water between alternative uses in accordance with the laws of economic efficiency. These laws only consider the direct use value for human consumption and the value of economic goods produced when water is used as an input. However, supporting water allocation between alternative uses based on laws of economic efficiency can lead to socially inefficient allocation, especially when the choice is between human consumption and agricultural or industrial uses. Such market mediation may therefore give rise to the violation of the individual's right to water.

From this perspective, and given water's crucial role in human survival, it is perfectly admissible for society to establish priorities. In this context, when water supplies are insufficient, it seems quite acceptable that priority should be given to direct human consumption over other uses, such as leisure or even over irrigation. Although agriculture is vital for guaranteeing other human rights, like access to food, it is possible to farm without irrigation, whereas it is impossible for a human being to survive without drinking water. However, in many parts of the planet, mainly in developing countries, the lack of access to irrigation water can lead to a denial of the right to food and indirectly to a violation of the right to water in those cases where water is indispensable for crop production.

As seen before, water is in its essence a common pool good. This classification is especially appropriate when the alternative uses of water are prioritized. In order to optimise the management of common pool goods, mainstream economics can resort to game theory, namely to the prisoner's dilemma. According to this game, the best outcome for each individual user is to act selfishly while the other user acts cooperatively and the worst outcome is to act cooperatively while the other user acts selfishly. The best outcome for society and resource conservation, the one which best ensures the human right to water, is that in which there is cooperation among the multiple users of the resource. However, this cooperation is only possible provided the existence of strong public or community commitment. This outcome, if used to select between alternative uses or in the allocation between users, maximizes social welfare but requires a mediation that is beyond the competitive market mechanism.

In order to minimise the conflicts emerging from the sharing of common pool resources, user participation in the decision-making process about resource allocation should be encouraged. Material and moral sanctions which threaten users' reputations can serve to penalise any selfish individual actions. With respect to user participation and resource valuation, group valuation proposed by social and political theory has also gained increased attention recently. This valuation method is based on the principles of deliberative democracy and assumes that public decision-making should result from an open public debate rather than from the aggregation of separately measured individual preferences (De Groot, 2006).

Social outcome validation by game theory and group valuation have been implemented for quite a while in many rural and urban communities in the management of common pool resources. The *Tribunal de las Águas de Valência* is a good example of an ancient communitarian institution with a one thousand year old history of allocating water. In this case, both economic and political theory and ancestral experience indicate the same path, questioning the need for alternative solutions, such as the market, to allocate water use.

Finally, the market is ecologically inefficient. Indeed, water is exhaustible over a given period of time, which means that its use can only be renewable if the extraction rate is lower, or equal, to the recharge rate. This is a vital question, since ensuring human rights in general and the right to water in particular does not contemplate any point in time beyond which it would be acceptable for a human right to cease to be guaranteed. This implies that water use must be sustainable. It is therefore difficult to conciliate individual and social interests, since no market mechanism can prevent individual consumption from exceeding the recharge rate. This issue becomes more acute in case of water contamination as the price mechanism per se cannot generate sufficient incentives for water treatment and reuse by the market, except in regions where extreme scarcity boosts water prices.

The water market is not competitive

Another reason for arguing against market control of the right to water stems from the fact that real water markets do not resemble the competitive market model (Henriques et al., 2006). First of all, water presents physical characteristics such as mobility and property of state change which require its transportation and delivery as a flow rather than as stock. Measuring and monitoring this flow is both complex and costly, which can frequently become an obstacle to determining clear property and usufructuary rights.

Second, water is an irreplaceable and indispensable good and therefore cannot be appropriated in an exclusive regime by any given user or supplier. Third, the almost exclusive primary producer is nature which by definition does not behave as an economic agent. Fourth, for many communities water is a free gift of nature, and there are strong social, cultural and religious objections to establishing a price and consequently a market for water. Fifth, competitive market mechanisms are unlikely to be able to meet the demand for non-consumptive market uses, such as recreational and ecological ones. Sixth, most consumptive water uses have side effects, such as pollution, which should be dealt with within the framework of the public water supply and be subject to public control.

Finally, for technological reasons, water distribution can be considered a natural monopoly in the sense that even if competition for concessions is allowed, the consumer cannot choose his supplier as he can, for instance, with cable or telephone. If one is dissatisfied with one's cable or telephone supplier, one can change. On the contrary, an individual cannot switch his water supplier. Therefore, if economic, social and cultural rights are to be taken seriously, it is of the utmost importance that decisions concerning water distribution should be made by all those affected by that same distribution, which means that it should be submitted to democratic (i.e. public) control, or at the very least mixed, rather than solely private management.

Conclusion

The main conclusion of this paper is that the market is, indeed, neither able to meet the needs and rights of individuals nor those of the planet and is, therefore, not qualified to secure the human right to water in Africa, or anywhere else. We have shown that there is an intrinsic conflict between the logic of the market and that of the provision of goods as human rights. In the first place, one must stress the normative character of this particular sort of provision. Indeed, to produce goods and services in order to satisfy a consumer's viable demand is one thing; to produce these exact same goods and services in order to satisfy a citizen's request is quite another. The tension lies precisely along the line that separates these two different manifestations of an individual need; the former is private and therefore economic in nature, whilst the latter is public and therefore political. By making the market an absolute value and an infallible means of rationally allocating all goods, the community runs the risk of reducing all categories of goods, and thus of rights, to the nature of the commodity. Now, this *commodification* of society is contradictory to the ideal situation in which society chooses to make a priority of enhancing human rights. Such a society prizes accountability and universality; market ideology happens to ignore both. Further to these two aspects, the market economy applied to water supply also turns out to be both inefficient and uncompetitive.

There is little practical evidence in Africa that privatisation does in fact result in increased efficiency, economic growth and development. Private firms are interested in profit not in social objectives whilst end users are unable to pay the tariffs required by privatised companies. The impression one gets is that in Africa, as in many parts of the developing world, privatization of water has not lived up to expectations. This market failure should prompt African governments as well as donor countries, intergovernmental institutions and development agencies to assist the State in guaranteeing the right to water. The dogmatic injunction that all services that *can* be privatized *should* be privatized must, therefore, be substituted by a set of policies prioritising human rights. One must bear in mind that such rights are not optional but binding governmental obligations which must be universally enjoyed. This demand also requires a different approach to economics. Indeed, one must take into consideration that the non-compliance of the market economy with regard to human rights is not the result of erroneously carrying out the right economics, but of correctly implementing the wrong economics.

References

Arrojo, Pedro. (2006). *El Reto Ético da la Nueva Cultura del Água: Funciones, Valores y Derechos en Juego*. Barcelona: Ediciones Paidós.

Bayliss, Kate and Hall, David (2000). *Privatisation of Water and Energy in Africa*. London: PSIRU, University of Greenwich.

Conteh, Sonkita (2006). *Inhibiting "Progressive Realization"? The effect of Privatization on the Right to Water in Senegal and South Africa*, LLM Dissertation, Faculty of Law, Centre for Human Rights, University of Pretoria.

De Groot, Rudolf S. (2006). Function Analysis and Valuation as a Tool to Assess Land Use Conflitcs in Planning for Sustainable, Multi-Functional Landscapes. *Landscape and Urban Planning* 75: 175–186.

Henriques, P., Branco, M., Fragoso, R. and Carvalho, M. L. (2006). Direito de Acesso à Água: Princípios Económicos para o seu Usufruto na Agricultura. In Branco, Manuel, Carvalho, Maria Leonor and Rêgo, Conceição (eds), *Economia com Compromisso: Ensaios em Memória de José Dias Sena*, p. 29–55. Évora: Universidade de Évora, Cefag-UE.

2003 IYFW (2008). A look at water resources in Africa, <www.wateryear2003.org> (accessed 5 January 2009).

Maltz, Hélio (2005). Porto Alegre's Water: Public and for All. In Balanyá, Belén, Brennan, Brid, Hoedeman, Olivier, Kishimoto, Satoko and Terhorst, Philipp (eds), *Reclaiming Public Water Achievements, Struggles and Visions from Around the World*, p. 29–36. Amsterdam: TNI/CEO.

Offenheiser, Raymond and Holcombe, Susan (2003). Challenges and Opportunities in Implementing a Rights-Based Approach to Development: An Oxfam America Perspective. *Nonprofit and Voluntary Sector Quarterly* 32 (2): 268–306.

Petrella, Riccardo (2004). *Désir d'Humanité. Le Droit de Rêver*. Bruxelles: Éditions Labor.

UN (2002). Substantive Issues Arising in the Implementation of the International Covenant on Economic, Social and Cultural Rights. General Comment N° 15. Committee on Economic, Social and Cultural Rights, 19–29 November, <www.unhchr.ch/html/menu2/6/gc15.doc> (accessed 4 April 2007).

UN (1997). *Comprehensive Assessment of the Freshwater Resources of the World*. New York: Commission on Sustainable Development.

UNDP (2006). *Human Development Report. Beyond scarcity: Power, poverty and the global water crisis*, hdr.undp.org/hdr2006/pdfs (accessed 10 May 2008).

UNECA – UN Economic Commission for Africa (2006). *Water in Africa – Management Options to enhance Survival and Growth*, NEPAD.

UNEP (2002). GEO3. The State of the Environment and Policy Retrospective: 1972–2002. In *Global Environmental Outlook*, <www.unep.org> (accessed 5 Sep 2008).

WHO (2000). *The Global Water Supply and Sanitation Assessment 2000*. Geneva.

WWF (2002). The Facts on Water in Africa, <www.panda.org/livingwaters> (accessed 10 October 2008).

Paula Duarte Lopes[*]

Implementation of the human right to water: Multi-level challenges

Abstract. The debate on whether the Human Right to Water exists or should exist is currently on the international agenda. The global acknowledgement of the urgency of reaching a decision on this matter is not an isolated issue, since national actors have been extremely active on this matter since the 1990s. We argue herein that, although it is of utmost importance to reach a consensus on the international consecration of the Human Right to Water, in practice, there is a very good chance that such recognition will have limited implications. Africa constitutes a clear example of this situation.
Keywords. Human right to water, international governance, domestic governance, local practices

Introduction

The debate on whether the Human Right to Water exists or should exist is currently on the international agenda. Since the 1990s, this question has increasingly gained momentum resulting, recently, in the creation of a United Nations Expert Committee on the Human Right to Water and Sanitation (UNHRC, 2008).[1] This global acknowledgment of the urgency

* Faculdade de Economia da Universidade de Coimbra. Av. Dias da Silva, 165. 3004–512 Coimbra – Portugal. Email: pdl@fe.uc.pt.

1 Although the United Nations decided to address the issue of water and sanitation together, the discussion in most fora is still presented in terms of water. Consequently, this paper will address the issue of a Human Right to Water, although the author

of this matter is not an isolated issue, since national actors have been extremely active on this matter since the 1990s, with the ultimate goal of obtaining legal recognition of the Human Right to Water.

We argue herein that, although it is of utmost importance to reach a consensus on the international consecration of the Human Right to Water, in practice, it is unlikely that such recognition will generate significant increases in the number of people with access to potable water and connected to some type of sanitation system as established by the Millennium Development Goals (Objective 7, Target 3)(UNGA, 2000). Africa constitutes a clear example of this situation.

Three sections address the timing of the debate, the dynamics underway and their implications. We start by analysing the context in which the Human Right to Water emerged and how it gained a generalized sense of urgency both nationally and internationally. It is argued that the timing of these efforts resulted from the marketisation of the water sector which has been taking place since the early 1990s. The pressure to commodify water resources along with the promotion of private sector participation has come to challenge the universal provision principle that had previously characterized water governance. The first section of this paper analyses exactly this change in water governance models and its implications for access, starting by contextualizing world water from a human perspective. It is argued that until the 1990s, a 'hydraulic state' water governance model was in effect, within which the Human Right to Water is implicitly assumed. This governance model is challenged throughout the 1990s with the consolidation of a 'neo-liberal market' water governance model, resulting in the international acknowledgement of the economic value of water and, due to its ill-considered implementation, in the violation of the universal provision principle. This shift triggered the debate, demands, efforts, lobbying for the consecration of the Human Right to Water.

The following section describes such efforts at the national, international and local levels, addressing not only the legal consecration of the Human Right to Water (or related efforts) as well as local practices protecting together with multi-level activism claiming the Human Right

---

recognizes that in several countries both services are provided by the same organization and charged in the same bill.

to Water. The dynamics are illustrated with specific international, African (regional), national and local cases.

This multi-level mapping of the efforts towards the consecration and implementation of the Human Right to Water constitutes a first step towards the analysis of their implications concerning the actual results in terms of increased number of population with access to drinking water (so what?). The last part of this section illustrates its argument with data from the African situation. The last section offers some final conclusions and identifies new avenues of research and action on the topic.

Why now?

Before analysing the evolution of water governance models and their implications, it is important to provide an overall picture of the situation concerning worldwide water resource availability and use. The quantity of water on Earth is constant and there is more than enough for human consumption. Nevertheless, fresh water resources constitute a small portion of the water available on Earth: only 2.5% of Earth's water resources are fresh (Shiklomanov, 2000). Moreover, of these, only 0.26% is readily available for human retrieval in lakes, rivers and easily accessible aquifers (Shiklomanov, 2000). The rest is either frozen at the poles or too difficult or expensive to pump from underground sources (Shiklomanov, 2000). Despite the constant quantity of water on the planet, water use patterns have changed over time and differ across regions. During the 20th century, use changed dramatically, with an unprecedented increase in water demands due to the industrial and agricultural revolutions and population growth. Nevertheless, "agriculture still takes the lion's share" (HDR, 2006: 137), being responsible, between 1998 and 2002, for more than 40% of water consumption in high-income countries of the Organisation for Economic Co-operation and Development and for more than 80% in developing countries (HDR, 2006: 138).

This has led to enormous pressure on water resources and given rise to fierce conflicts among users. Adding to this dilemma is the fact that water is often in the wrong place, at the wrong time, in the wrong quality. In 2002, the World Health Organization "estimated that 1.1 billion people (17% of the global population) lacked access to improved water resources, and 2.6 billion people (42% of the global population) lacked access to improved sanitation" (Hutten & Haller *in* Dubreuil, 2006: 3) "which is the primary cause of water contamination and diseases linked to water" (ECOSOC, 2003: §1). At the beginning of the 21st century, these figures are alarming. Governing water resources has moved beyond the traditional technical realm and gained unprecedented prominence both socially and politically.

Water governance

Until the late 1980s, water resources were governed by a hydraulic state model, which was characterized by public ownership of water resources, the principle that universal provision of water and sanitation services was governments' responsibility, and a hydraulic supply-led approach to water management. Water's public ownership was often assumed under national legislation as part of the natural soil and subsoil resources. Most countries did not have water's public ownership explicitly enshrined in legal terms or specific water legislation. The state was, however, considered the designated and default entity responsible for water management, since water was public property. Furthermore, every state (at different levels) included in its governing manifesto the goal of expanding the water and sanitation networks in order to cover 100% of the population. Governments adopted a pragmatic managerial approach to water resources, which translated into large-scale infrastructure projects – dams, water diversions, hydropower plants, and irrigation schemes – to take and guarantee water where and when it was needed. States mobilized water "as a strategic resource for societies", as a means for supporting modernization, industrialization, urbanization, and agricultural revolution processes (Bakker, 2003: 19).

In economic terms, this basically meant that water resources were considered as a social good.[2] This was the result of a combination of

2 Social goods are political public goods, which may be delivered as economic private goods, since their consumption is excludable (potential consumers can be excluded

things: perception of abundance[3] (leading to the assumption of water as an economic public good), its vital role for human life, and non-substitutability (political public good, i.e., social good). Although water governance varied depending on each country's history, culture and water availability, overall, "there has been an evident tendency towards [the] socialization of water resources" (Popov, 2001: 2). Consequently, water's allocation was not based on market prices, since the state maintained water prices (when they existed) artificially low in order to guarantee universal provision. The allocation mechanism was based on need (state guarantees) and not on consumption (consumer pays).[4]

In the 1970s and 1980s, three combined developments set in motion the creation of a new water governance model. Firstly, from the 1970s onwards, the costs of the hydraulic state model began to receive greater attention. On the one hand, there were financial considerations The cost-effectiveness of hydro-electric dams came under question. Many industrialized countries witnessed a stagnation of water demand and most governments faced deep fiscal constraints, in the wake of the oil crises (Bakker, 2003: 23–24). On the other hand, "years of lobbying by conservationists and environmental groups began to change public consciousness" (Bakker, 2003: 23). Citizens became more aware of the costs of the hydraulic state model including the loss of animal species, especially aquatic ones; the disruption of ecosystems; the flooding of cultural and religious sites; the displacement of populations; and environmental degradation (Gleick, 2000). They began to demand a more environmentally conscious approach.

---

from access) and rival (one's consumption of water leaves less water available for others). They are, however, usually delivered by public entities and financed by public funds, due to social policy concerns and objectives. See Lopes, 2006.

3 Water is technically a renewable natural resource; consequently, the perception that it is abundant and therefore its consumption non-rival and non-excludable (economic public good) was (and still is in some places) assumed.

4 There are two major exceptions to this approach dating from the nineteenth century: the United States of America and France. For further information on these cases see, among others, Water Science and Technology Board, 2002, Gleick et al, 2002 and La Documentation Française.

Simultaneously, an economic paradigm shift took place in the late 1970s. From a policy paradigm where government was part of the solution, Neo-liberal economics claimed government was part of the problem. The neo-liberal economic paradigm has (convincingly) presented private enterprise as "one of the most desirable methods of allocating resources, goods, and services" (Trelease, 1965: 6). This desirability follows from the belief that "exclusive private property is thought to foster the well-being of the community, giving its members a medium in which resources are used, conserved and exchanged to their greatest advantage" (Rose, 1986: 712). Consequently, resource allocation should unfold under the auspices of the market.

Thirdly, in January 1992, the World Meteorological Organization convened the International Conference on Water and the Environment, leading to the so-called Dublin Principles, presented that same year at the United Nations Conference on Environment and Development in Rio de Janeiro. The Dublin Principles have become extraordinarily influential. The first principle acknowledges the vulnerable and finite nature of water resources, and recognizes its role in sustaining life, the environment and development. The central role of women and the need for a democratic and participatory management model are asserted in the second and third principles. But it is the fourth principle that has been the most cited and used to support new water governance options.

Principle No. 4 – Water has an economic value in all its competing uses and should be recognized as an economic good (ICWE, 1992).

As a result of these three developments, governments began to search for ways to both financially discipline and improve efficiency in the water sector, resorting to a marketization process.[5] The immediate consequences were two-fold. Many governments decided to allow private entities to participate in the water sector, expanding the market opportunities for private investment and directly questioning the public nature of this sector. Water charges constitute the other prong of marketisation. Prices became the preferred mechanism for achieving efficiency and, allegedly, sustainability. Marketisation is, therefore, at the heart of the neo-liberal

5 The author adopts the term marketization instead of privatization. For information on the underlying arguments see Lopes, 2006, 2009.

market water governance model, where private sector participation and commodification constitute the operative instruments in order to promote efficiency and economic sustainability.

From social to economic equity

The neo-liberal market water governance model challenged the hydraulic state one at its core. This challenge is based on two main modifications: economic instead of social equity and state instead of market failure. In the first case, the new model is characterized by economic equity facilitated by the market. Consequently, the objective of universal provision and the principle of public service and public nature applied previously to the water sector are structurally changed. The goal becomes recovering costs and improving efficiency and the governing principles are market oriented. Any failures to provide water are defined as failures of the state, rather than the market. The water sector was, consequently, opened to private sector exploitation with a subsequent reduction in public spending on the water sector together with a growing public perception that the states were exonerating themselves from their responsabilities regarding water governance.

Water resources and services have always been political assets, but, unlike other natural resources, they have consistently resisted transformation into economically tradable assets. Although, it is clearly understood, from an economic point of view, that these are economic goods and services (Köllicker, 2004), from a political point of view, they have been classified as essential social goods, implying the need for universal provision. Consequently, marketisation, without any previous re-negotiation of social expectations or traditional citizens' rights, resulted in a modification of the basic conditions for accessing water resources and services.

Reactions to these changes around the world have ranged from violent social unrest (in Bolivia) to indifference (in Portugal), from frantic legislative dynamics (in Canada) to campaigns in favour of The Right to Water (in Mali) and Sanitation (e.g. United Nations). In fact, the universal provision principle of the hydraulic state governance model in conjunction with the tendency for public ownership and management of water resources and services have created the perception of a Human Right to Water. The state is expected to be the pre-eminent promoter and guarantor of basic Human Rights in a country. Thus, the hydraulic state governance model implicitly acknowledged water as a Human Right. The neo-liberal market water governance model not only challenged the hydraulic state one at its core, as explained above, but it explicitly disregarded the Human Right to Water, by defining access in terms of prices and allowing for exclusion based on economic equity, without consideration for social concerns.

What is being done?

This shift has placed the water sector under enormous stress. Reactions have varied across time and place. In order to discuss their implications, it is essential to first identify the different initiatives which have taken place concerning the Human Right to Water. These dynamics unfold at different levels – nationally, internationally, and locally. Although for each case, and at the various levels, unique events have triggered the discussion regarding the Human Right to Water, they are all a result of the broader change analysed in the previous section: the marketisation of water resources – private sector participation and commodification – which violated the universal provision principle.

National Constitutions

Nationally, water marketisation led to political resistance and social mobilisation across the world. The perception that the state was retreating not only triggered opposition from international organisations, such as the Association for a World Water Contract (ACME), but has also led to domestic social protest from supporters of public water systems as well as anti-globalisation and anti-free trade activists. In several countries, such as Argentina, Bolivia, The Philippines, Great Britain (England and Wales), the United States of America, and Uruguay, citizens have reacted against private sector participation, against water charges, and against what they understood to be the reneging of governments on their obligations and responsibilities. They demand 'the return of the state' and the removal of private actors from the water sector. These demands have materialised in the form of the termination and suspension of public-private contracts, revocation of legislation passed under the neo-liberal market movement, and new legislation concerned with the universal provision of water not guaranteed by the market.

In some cases, violence ensued (e.g. Bolivia), whereas in others non-violent means were adopted, such as holding a *referendum* on the Human Right to Water (e.g. Uruguay) or holding a parliamentary discussion on the issue (e.g. Belgium). Although in several of these cases (e.g. Bolivia and Uruguay) national legislation, eventually, came to recognise the Human Right to Water, several other countries have legally recognised the Human Right to Water without outbursts of violence or social mobilisation. Still, other governments, for various reasons, fell short of recognising the Human Right to Water, but reinforced the public nature of water resources within their borders, by clarifying water's legal and economic status and by increasing regulation on water use conditions (e.g. Canada).

The news of political discontent regarding challenges to the universal provision of water in conjunction with the unprecedented situation some governments have faced led them to adopt specific legislation concerning water resources. For example, in Africa, since the mid-1990s, several countries have included the Human Right to Water in their constitutions: Uganda and Ethiopia (1995), South Africa, Zambia and Gambia (1996) and Congo (2006). South Africa explicitly recognises that "everyone has the right to have access [...] to sufficient food and water" (1996, Art. 27, 1.b.). In the Republic Democratic of Congo the right to access potable water is guaranteed (2006, Art. 48). In Uganda, the Constitution states that the State "shall endeavour to fulfil the fundamental rights of all Ugandans [...], in particular, [...] rights and opportunities and access to [...] clean and safe water", among others (1995, XIV, b.). Ethiopia, Zambia and Gambia do not explicitly recognise the Human Right to water, but they do recognise the responsibility of their states to provide or facilitate access to clean and safe water (1995, Art. 90, 1.; 1996, Art. 112, d.; 1996, Art. 216, 4., respectively).

It is undisputable that the legal-constitutional efforts several countries have dedicated in order to enshrine the Human Right to Water reflect a concept of public service encompassing the hydraulic state governance model and preoccupations with the emergence of the neo-liberal market one. It should be noted, however, that in all these cases, although the Human Right to Water has been explicitly or implicitly consecrated, the possibility for private sector participation in water governance has not been not addressed. In fact, Uruguay is the only country that has legally established the exclusion of private sector participation in the water sector, as a result of the 2004 *referendum*. The results determined that water should be considered a Human Right and that water governance should be kept exclusively in the public sphere.

International dynamics

Internationally, one can identify different dynamics. Since the adoption of the Universal Declaration of Human Rights (UNGA, 1948), most actors have looked towards the United Nations for guidance. It can be argued that the Human Right to Water already exists within the Declaration, since "[s]atisfying the standards of Article 25 [right to food, housing and health] cannot be carried out without water of sufficient quantity and quality to maintain human health and well-being" (Gleick, 2000). Although history has shown that the provision of these basic rights is the prerogative of the each state, governments' policies and practices throughout history have created a perception and expectation by their citizens that water is, in fact, a Human Right.

Consequently, until the 1980s, the Human Right to Water was allegedly assumed within the general rights to life, food and health as well as in the International Covenant on Civil and Political Rights (UNGA, 1966), and in several regional treaties and conventions on the protection of human rights. It was, however, explicitly addressed in the Geneva Convention (1949: Treatment of Prisoners of War and Protection of Civilian Persons in Times of War) and in the Additional Protocols (1977: Protection of Victims of International Armed Conflict and Protection of Victims of Non-international Armed Conflict). All of these conventions and protocols have established the right to sufficient food and drinking water and the prohibition of destroying drinking water installations in situations of armed conflict. Consequently, it can be assumed that the Human Right to Water exists in times of peace as well. Additionally, two other specific conventions, explicitly establish the right to drinking water: The Convention on the Elimination of Discrimination Against Women (1979, Art. 14.2) and The Convention on the Rights of the Child (1989, Art. 24.1). Again, it can be concluded that if this issue is explicitly referred to in relation to discriminated or at-risk groups, it is assumed that the Human Right to Water exists for all inhabitants.

Probably the most important response, so far, to the challenges to the Human Right to Water has been the General Comment No. 15 by the UN Committee on Economic, Social and Cultural Rights: "The human right to water entitles everyone to sufficient, safe, acceptable, physically accessible and affordable water for personal and domestic use" (ECOSOC, 2003). Although this General Comment does not constitute a treaty in and of itself, "it is based [...] on the general acceptance of fundamental human rights like the right to life and the right to health" (Dubreuil, 2006: 8). This Comment has undoubtedly contributed to the international momentum towards the recognition of the Human Right to Water "by offering considerable legal and moral support to the global efforts aimed at attaining the Millennium Development Goal related to water" (Salman and McInerney-Lankford, 2004: 88).

More recently, the United Nations gave yet another sign that it is seriously considering the issue of the Human Right to Water by nominating an Independent Expert on the Issue of Human Rights Obligations Related to Access to Safe Drinking Water and Sanitation (UNHRC, 2008) with a three year mandate. This was a result of the Report presented by the High Commissioner for Human Rights in 2007, which stated

> that it [was then] time to consider access to safe drinking water and sanitation as a human right, defined as the right to equal and non-discriminatory access to a sufficient amount of safe drinking water for personal and domestic uses [...] to sustain life and health. (UNHRC, 2007)

6 Both the World Water Council (1996) and the Global Water Partnership (1996) were created in order to influence water governance, namely through the socialisation of

At the same time, other international entities have addressed this issue. Anti-free trade and anti-globalization movements, as well as international water-specific organizations, have joined in the fight against water marketisation. They oppose the initiatives developed by the World Water Council, the Global Water Partnership,[6] the United Nations, and multinational water companies. The water-specific members of this group oppose the commodification of water, endorsing water as a Human Right. This group is basically a coalition of small non-governmental organisations, which work "on development, health, human rights, education and environmental issues, together with a very significant presence of public sector unions" (Morgan, 2004: 2).

One of the members of this network has gained some public awareness, especially in Europe: the Association for a World Water Contract–France (ACME).[7] They have, in their own words, "no other legitimacy or representativeness than that of being [concerned] citizens" from all over the world (ACME, 1998). The ACME goes further than just opposing the commodification of water. It claims water is a *res publica* on a planetary scale and should be governed as such. Moreover, they argue that the access to water is an "inalienable individual and collective right" (Principle 2) (ACME, 1998; Touly, 2004). On the financial side, the ACMEs recognize that there are costs, but they argue those costs should be borne by the collective as a whole.

> The financing must be ensured by collective redistribution. The mechanisms of individual price-fixing [...] must start from a level of water usage that goes beyond the vital and indispensable minimum. Beyond the vital minimum, progressive pricing must be a function of the quantity used. (ACME, 1998)

---

the Dublin Principles and the adoption of integrated water management. Their main goals are to promote a water secure world through raising awareness and promoting sustainability. The World Water Council organises the World Water Forum every three years. For more information check their website: <www.worldwatercouncil.org/> and <www.gwpforum.org>.

7 The ACME was created in 1998 and was a result of the mobilisation of several eminent persons, such as Mário Soares, Riccardo Petrella, Danielle Mitterrand, Susan George, Vandana Shiva, Le Prince Laurent de Bélgique, and a number of non-governmental organizations. Several countries have created their own ACMEs. For more information check their website: <www.acme-eau.com/index.php>, accessed February 10, 2005.

The ACME as well as other players, including several prominent academics and researchers, have been advocating the establishment of a guaranteed minimum quantity of water per person per day for human consumption to be delivered free of charge to each and every human being on the planet as a means to guaranteeing the Human Right to Water (universal provision). Different figures have been put forward, but a consensus seems to have emerged associated with a quantity between 20 (defended, for instance, by the World Health Organisation on its website) and 50 litres (Dubreuil, 2006: 3) of water per person per day.[8]

The World Water Forum (organized every three years by the World Water Council) has mostly focused on creating a ministerial consensus on water governance, supported by the insight of multinational water companies and several private consultancies. Although the World Water Council supports the recognition of the Human Right to Water, it stresses the arguement for, as in the Dublin Principles, "affordable" as opposed to free water and makes a case for the "allocation of about 20 litres of water per day" for each individual (WWC, n/a). It was not until the 5th World Water Forum (2009) that the Human Right to Water gained widespread visibility, with different state, municipal and civil society representatives arguing for a common ministerial position. Although consensus has not yet been reached, the issue has been established as a major topic of debate on the international agenda for the coming years.

African references

Regional as well as international debate on the Right to Water issue has been noticeable and the African case is no exception. The African Charter on Human and People's Rights (OUA, 1982) does not address the issue explicitly, similarly to the Universal Declaration of Human Rights (UNGA, 1948). Nevertheless, the African Charter on the Rights and Welfare of the Child explicitly states that "State Parties [...] shall take measures [...] to ensure the provision of adequate nutrition and safe drinking water" (OAU, 1990: Art. 14, 2.c.). Consequently, as discussed above, the same reasoning

8 The ACME advocates a minimum of 40 litres (Mitterrand, 2006).

applies: if this right is explicitly established in relation to children, then, it is reasonable to assume that the Human Right to Water also exists in relation to the overall population.

Additionally, the decision taken by the African Commission on Human and People's Rights on "Free Legal Assistance Group and others v. Zaire" again contributes to the recognition of the Human Right to Water. The Commission found that failure by the government of Zaire "to provide basic services such as safe drinking water [...] constitutes a violation" (ACHPR, 1995: § 47) of each individual's "right to enjoy the best attainable state of physical and mental health" (OUA, 1982: Art. 16, 1.). This decision creates a precedent confirming states' responsibilities in meeting the Human Right to Water.

Local practices

Finally, at the local level, the dynamics have been quite different and much more silent. In most countries the actual locus of power concerning water distribution resides at the regional or local level with public, or public-private, water institutions that manage daily access conditions for the public in general. It is at this level of implementation that access is or is not provided, ensuring or not the Human Right to Water. The focus is, firstly, on how to secure access to the water supply system and, secondly, on what the conditions are to maintain access. In other words, it is the daily management of these water institutions that in the end determine if one has access to water and under what conditions.

In certain Portuguese municipalities, families can avail themselves of water relief from the municipal government, to avoid having their water supply disconnected, if they are unable to pay.[9] Families can apply to the municipality, and, based on their economic situation, the municipality may decide or not to pay for their water bill. This practice is not based on constitutional or national legislation, but rather on a municipal recognition

9 This information was obtained from a discussion at the International Seminar "Em Defesa do Direito à Água" [Defending the Right to Water] that took place in Lisbon on January 10, 2009, at the Instituto de Ciências Sociais.

that disconnecting water based on a lack of funds is a community-sensitive and moral issue.

In other cases, although the Human Right to Water is constitutionally recognised (1996), and although governmental efforts have been developed towards the guarantee of such right (see, for instance, Quinn and Marriot, 2006), it has been extremely costly and complex to ensure the Human Right to Water to each and every South African. Consequently, attempts have been made to transfer the impact of not paying the water bill to a different utility. For instance, if a family does not pay the water bill, it might have its electricity supply disconnected for a certain period of time, but not the water supply.[10]

These are merely illustrative examples of attempts to secure the Human Right to Water at the lowest levels, taking into consideration the actual situation of individuals. More research is needed to identify the situation and the mechanism where these practices are taking place. Nevertheless, these two examples, albeit requiring further substantiation, highlight the relevance of including local practices in the analysis of the implementation of the Human Right to Water. They further alert to the financing difficulties on the one hand and the need for ingenuity on the other when addressing the issue.

So what?

The recognition of the Human Right to Water constitutes a first step towards guaranteeing sufficient water for each human being's personal and domestic consumption. To this end, the international efforts underway are crucial for helping states to commit, increasing the roll of those who

10 This information was provided by Michela Marcatelli (Institute for Social Studies, The Hague, The Netherlands), participant on the panel "Water as a Human Right" at the International Conference on 'Water in Africa: Hydro-Pessimism or Hydro-Optimism?', held in October 2–3, 2008, organized by the Centre for African Studies at the University of Porto, Portugal. Her presentation was on "Access to Water in Post Apartheid South Africa: Right or Commodity?".

have already recognised the Human Right to Water in their constitutions. It should be noted, however, that despite the current controversies, recognising the Human Right to Water is actually not the difficult part. The difficulties arise, as recognised by several institutions that have vetoed the consecration of the Human Right to Water, regarding the implications of implementing such right. The state, as mentioned above, is the primary promoter and guarantor of human rights. A human right imposes three intertwined obligations: respect, protection and fulfilment (Dubreuil, 2006: 8; Sacher and Windfuhr, 2008).

The respect for the Human Right to Water implies that governments must recognise the right of each inhabitant to "sufficient, safe, acceptable, physically accessible and affordable water for personal and domestic use" (ECOSOC, 2003). This means existing access to water should not be destroyed. Individuals or groups should not be arbitrarily disconnected, even in case of default due to lack of funds and there should be no discrimination regarding access to water (Sacher and Windfuhr, 2008: 164). But this recognition means nothing if the right is not protected and enforced.

The obligation to protect entails controlling all parties involved in water politics, public and private, in order to prevent any violation of the Human Right to Water (Sacher and Windfuhr, 2008: 164). This requires an active engagement in overseeing the fulfilment of the right and in applying sanctions on the non-abiding. Fulfilling the Human Right to Water is extremely challenging, since it requires pro-active management and investment policies geared towards the expansion of the national water supply system coverage, the guarantee of a minimum quantity of water delivered to each and every inhabitant (al Jayyousi, 2008: 130), and the assurance of a continuous supply of a quantity and quality of water matching national needs. This requires coordinated and coherent policy-making among the different relevant policy areas and institutions (Sacher and Windfuhr, 2008: 164).

As suggested above, guaranteeing a minimum quantity of water to each inhabitant may be achieved, at the local level, without the recognition of the Human Right to Water. However, if expansion or maintenance of the water supply system is needed or if water availability decreases below a sustainable level, local authorities may lack the expertise, and/

or the political and financial means to address the problem. The respect, protection and guarantee of the Human Right to Water constitute a multi-level endeavour.

Although international (and domestic) legal frameworks consecrating the Human Right to Water would constitute a crucial step toward its promotion and guarantee, the implementation process is structurally dependent upon the policies and budgets of individual states as well as the know-how of local authorities. The state remains a key actor in implementing the Human Right to Water, since it is at the state level that national water policies are defined, financed and rendered operational. States may, however, fall short of guaranteeing the Human Right to Water due to a lack of funds or a shortage of water itself. It is these cases in which a legal guarantee becomes essential since such a statute would neither allow the issue to be left to national agendas nor permit states to renege on their responsibilities. The binding legal frameworks would provide the basis for ensuring the right to water as a fundamental right independently of changes in governmental policies or governing conditions.

Consequently, recognising the Human Right to Water becomes critical to embedding and sustaining its implementation process. Local utilities are critical as they interact with the actual users in order to implement the strategy and policies defined. They cannot, however, guarantee the Human Right to Water on their own. A water governance model structurally committed to universal provision and based on social equity is crucial for achieving and sustaining the Human Right to Water.

Africa

Despite African regional and national efforts towards the recognition of the Human Right to Water, the situation is still dire. The "Millennium Development Goals: 2008 Progress Report" clearly illustrates the target of halving the African population without access to improved drinking water. Although in North Africa there is relatively high water coverage and in Sub-Saharan Africa there is low coverage, neither region has shown significant progress towards meeting the Millennium Development Goal in 2015 (UN, 2008a). It should be noted that Western Asia and the Asian

Countries of Commonwealth of Independent States have equally failed to make sufficient progress towards meeting their 2015 targets. However, at this moment it is the Sub-Saharan countries which not only have the lowest clean water coverage rates in the world (UNDP, 2006: 5) but also account for more than one third of the world's population without access to clean water (UN, 2008: 42).

Two additional factors challenge the assurance of the Human Right to Water, beyond its consecration. Firstly, Northern Africa is already facing water scarcity, with "more than 75 per cent" of its river flows being withdrawn (UN, 2008: 40), and "[a]lmost a quarter of Sub-Saharan Africa's population [is currently living] in a water-stressed country" (UNDP, 2006: 135–136).[11] These situations are not improving. On the contrary, with climate change they are actually deteriorating (see UNDP, 2006, 2007). It is estimated that by 2025, "90% of [Northern African] people will be living in water-scarce countries" and that average water availability in this region will be "just over 500 cubic metres per person" (UNDP, 2006: 136). As for Sub-Saharan Africa, it is forecast that "the share of the region's population [living] in water-stressed countries [will rise] [...] to 85% by 2025" (UNDP, 2006: 136).

Secondly, economic scarcity currently characterises most of Sub-Saharan Africa, "where [inadequate] human, institutional and financial capital limit[s] access to water", implying the "lack of or underdeveloped water infrastructure, high vulnerability to short- and long-term drought, and difficult access to reliable water supplies, especially for rural people" (UN, 2008: 40). Moreover, in this region, there has been insufficient progress toward reaching the Millennium Development Goal of halving the number of people living in extreme poverty (below $1 per day). This is the only region characterised by 'very high poverty' (UN, 2008a).

These figures, unfortunately, reflect the extreme relevance of the argument herein. The consecration of the Human Right to Water will have very limited impact on the actual number of people with access to drinkable water. It is crucial to introduce into the equation not only the different levels at which these efforts are taking place, but also the distinct aspects connected with the respect, protection and guarantee of the

11 Figures vary, but here it is assumed that the water-stress threshold is 1,700 cubic meters per person per year, and that the water-scarcity one is 1,000 cubic meters.

Human Right to Water, such as environmental, development, urbanisation, and agricultural policies, amongst others.

Conclusion

The world water situation is worrisome, and it is most dire in Africa. The consecration of the Human Right to Water is one of the efforts to address the situation at the individual level. Consecrating such a right, although extremely important, since it imprints urgency to the debate, will not *per se* reduce the number of people without access to safe and clean water nor will it guarantee the continuity of service for those currently with access.

It is essential to adopt a multi-level and multi-dimensional approach to water resource management. Efforts at all levels – international, regional, national and local – should be articulated in order to provide a unified front for the recognition and realisation of the Human Right to Water. Efforts in defining a water governance model, which include the Human Right to Water, should also be coordinated with decisions made in other policy areas. All other relevant policies should be embedded in water governance principles, establishing as a priority the universal provision principle based on social equity. It is essential that a sustainable approach is adopted and that funds are made available to finance the costs of respecting, protecting and enforcing the Human Right to Water. Water scarcity can only begin to be addressed if there is enough human ingenuity to overcome the situation. And human ingenuity requires the economic capacity to support such developments.

In conclusion, notwithstanding the shortcomings associated with the consecration of the Human Right to Water, its recognition would create international and national binding obligations, focus water policy on individual and community needs, and, allegedly, create the conditions for funds to be channelled towards its implementation.

References

ACHPR (1995). *Free Legal Assistance Group and others v. Zaire, African Commission on the Human and People's Rights*, Comm. No. 25/89, 47/90, 56/91, 100/93.

ACME (1998). Water Manifest, Association pour le Contrat Mondial de l'Eau, <www.f1boat.com/99/watermanifesto.html> (accessed 10 February 2005).

al Jayyousi, Odeh (2008). Water as a Human Right: Towards Civil Society Globalization. In Biswas, Asit K., Rached, Eglal and Tortajada, Cecilia (eds), *Water as a Human Right for the Middle East and North Africa*, p. 121–131. London, New York: Routledge.

Bakker, Karen J. (2003). *An Uncooperative Commodity: Privatizing Water in England and Wales.* Oxford: Oxford University Press.

Constitution de la République Démocratique du Congo (2006).

Constitution of the Federal Democratic Republic of Ethiopia (1995).

Constitution of the Republic of South Africa (1996).

Constitution of the Republic of The Gambia (1996).

Constitution of the Republic of Uganda (1995).

Constitution of Zambia (1996).

Convention (III) relative to the Treatment of Prisoners of War (1949).

Convention (IV) relative to the Protection of Civilian Persons in Time of War (1949).

Dubreuil, Céline (2006). *The Right to Water: From concept to implementation.* World Water Council.

ECOSOC (2003). Substantive Issues Arising in the Implementation of the International Covenant on Economic, Social and Cultural Rights – General Comment No. 15 (2002): The right to water (arts. 11 and 12 of the International Covenant on Economic, Social and Cultural Rights), United Nations Economic and Social Council, Committee on Economic, Social and Cultural Rights, E/C.12/2002/11.

Gleick, Peter H. (2000). The Human Right to Water. In Gleick, Peter H. (ed.), *The World's Water 2000–2001: The Biennial Report on Freshwater Resources*, p. 1–17. Washington, D.C.: Island Press.

Gleick, Peter H. (2000). The Changing Water Paradigm: A Look at Twenty-first Century Water Resources Development. *Water International*, 25 (1): 127–138.

Gleick, Peter H. et al (2002). The Privatization of Water and Water Systems. In Gleick, Peter H. (ed.), *The World's Water 2000–2001: The Biennial Report on Freshwater Resources*, p. 57–85. Washington, D.C.: Island Press.

ICWE (2002). The Dublin Statement on Water and Sustainable Development International Conference on Water and the Environment, <www.wmo.ch/web/homs/documents/english/icwedece.html> (accessed 10 February 2005).

Köllicker, A. (2004). *Globalization and National Incentives for Protecting Environmental Goods*, MPI Collective Goods Preprint No. 2004/3. Bonn, Germany: Max Planck Institute.

La Documentation Française (n/a), Les enterprises privées et la politique de l'eau, <www.vie-publique.fr/documents-vp/entrepriseprivee.shtml> (accessed 18 July 2005).

Lopes, Paula Duarte (2009). Sharing water: evolution, threats and challenges, *Lusotopie* 16 (1): 177–192.

Lopes, Paula Duarte (2006). *Water with Borders: Social Goods, the Market and Mobilization*, PhD Dissertation, Baltimore: The Johns Hopkins University.

Mitterrand, Danielle (2006). Communication at the 4th World Water Forum on 'Securing the Right to Water: from the Local to the Global, Civil Society Perspectives', Mexico City, March 19.

Morgan, Bronwen (2004). The Regulatory Face of the Human Right to Water, Paper presented at the 'Good Water Governance for People & Nature: What Roles for Law, Institutions, Science & Policy?', Dundee, Great Britain.

OUA (1990). *African Charter on the Rights and Welfare of the Child.* Organisation of African Unity: CAB/LEG/24.9/49.

OUA (1982). *African Charter on the Human and People's Rights.* Organisation of African Unity: CAB/LEG/67/3 rev. 5, 21 I.L.M. 58.

Popov, Dejan (2001). Consideration on Current Issues in Contemporary Water Law, Paper presented at the Regional Conference on 'Water Law: Legal Aspects of Sustainable Water Resources Management', Teslic, Republic of Servia, Bosnia and Herzegovina.

Protocol Additional to the Geneva Conventions of 12 August 1949, and relating to the Protection of Victims of International Armed Conflicts (Protocol I) (1977).

Protocol Additional to the Geneva Conventions of 12 August 1949, and relating to the Protection of Victims of Non-International Armed Conflicts (Protocol II) (1977).

Quinn, Nevil and Marriot, Susan (2006). Sustainable Water Resource Management in South Africa: A Decade of Progress? In Hill, Jennifer; Terry, Alan and Woodland, Wendy (eds), *Sustainable Development: National Aspirations, Local Implementation*, p. 67–90. Surrey: Ashgate Publishing.

Rose, Carol (1986). The Comedy of the Commons: Custom, Commerce, and Inherently Public Property. *University of Chicago Law Review* 53: 711–781.

Sacher, Danuta and Windfuhr, Michael (2008). The Debate on "Water as a Human Right" and Its Implications for Development Assistance. In Scheumann, Waltina; Neubert, Susanne and Kipping, Martin (eds), *Water Politics and Development Cooperation: Local Power Plays and Global Governance*, p. 147–176. Springer.

Salman, Salman M. A. and McInerney-Lankford, Siobhán (2004). *The human right to water: legal and policy dimensions.* Washington, D.C.: World Bank.

Shiklomanov, I. (2000). Appraisal and Assessment of World Water Resources, *Water International* 15 (1): 11–32.

Touly, Jean-Luc (2004). *Personal Communication* (Administrative Employee at *Général des Eaux* for 27 years; President of the ACME-France). Paris, France, July 16.

Trelease, Frank J. (1965). Policies for Water Law: Property Rights, Economic Forces, and Public Regulation. *Natural Resources Journal* 5 (1): 1–48.

UN (2008). *The Millennium Development Goals Report.* United Nations.

UN (2008a). *Millennium Development Goals: 2008 Progress Report.* United Nations.

UNDP (2007). *Human Development Report 2007/2008. Fighting climate change: Human solidarity in a divided world.* United Nations Development Programme.

UNDP (2006). *Human Development Report 2006. Beyond Scarcity: Power, poverty and the global water crisis.* United Nations Development Programme.

UNGA (2000). *United Nations Millennium Declaration,* United Nations General Assembly:

A/RES/55/2.

UNGA (1989). *Convention on the Rights of the Child.* United Nations General Assembly: A/RES/44/25.

UNGA (1979). *Convention on the Elimination of Discrimination Against Women.* United Nations General Assembly: A/RES/34/108.

UNGA (1966). *International Covenant on Civil and Political Rights.* United Nations General Assembly: 2200A (XXI).

UNGA (1948). *Universal Declaration of Human Rights.* United Nations General Assembly: 217A (III).

UNHRC (2008). *Independent Expert on the issue of human rights obligations related to access to safe drinking water and sanitation.* United National Human Rights Council: A/HRC/RES/7/22.

UNHRC (2007). *Report of the United Nations High Commissioner for Human Rights on the scope and content of the relevant human rights obligations related to equitable access to safe drinking water and sanitation under international human rights instruments.* United Nations Human Rights Council: A/HRC/6/3.

UNSG (2005). *Secretary-General Names Advisory Board on Water and Sanitation.* United Nations Secretary-General: SG/A/870.

Water Science and Technology Board (2002). *Privatization of Water Services in the United States: An Assessment of Issues and Experience.* Washington, D.C.: National Academy Press.

WWC (n/a). Right to Water: Moving towards a global consensus?, <www.worldwatercouncil.org/fileadmin/wwc/Programs/Right_to_Water/Pdf_doct/Story_RTW_CD_March07_compressed.pdf>.

PART II
Water governance and institutional issues

Theoretical reflections on the analysis of water governance in coupled social-ecological systems

Decentralization and struggles for water in Mali

Social learning and water resource management in Southern Africa

The Mphanda Nkuwa Project and China's impact
on Mozambique and Southern Africa's water development

Mathias Polak[*] and Stefan Liehr[**]

Theoretical reflections on the analysis of water governance in coupled social-ecological systems

Abstract. The Cuvelai basin in Northern Namibia and Southern Angola consists of a large number of small ephemeral watercourses flowing southwards from the Angolan highlands to the Etosha-pan. Especially in the Namibian part of this basin, water availability varies greatly, both seasonally and annually. The water supply of inhabitants is assured through the use of seasonal surface water, shallow hand-dug wells and an interbasin water transfer system from the Kunene river in Angola. Population growth and increasing herds of cattle threaten to push these supply patterns to their limits. Huge infrastructural developments together with institutional reforms under the concept of Integrated Water Resources Management (IWRM) have been undertaken since independence in order to ensure the reliable supply of water. The result is a fundamental change in water management from traditional forms of organisation to a system with mixed institutional patterns between state, market and community-based institutions.

This chapter analyses water supply and management from the perspective of resilience. This analytical concept deals with the ability of social-ecological systems to absorb external changes. Evolving natural conditions and anthropogenic factors challenge the traditional relations between humans and nature and require them to react. Our focus is on governance structures and therein on the possibilities of key players to steer the direction of institutional change. The paper brings together a significant combination of analytical tools from social ecology, resilience

* University of Frankfurt and Centre for African Studies, Porto. Lohmühlenstr. 22, Berlin; mathias.polak@web.de.

** Institute for Social-Ecological Research (ISOE). Hamburger Allee 45, Frankfurt; liehr@isoe.de.

research and the investigation of institutional change which should allow for the analysis of institutional change within social-ecological systems.
Keywords. Social ecology, Resilience, Water governance, Namibia, Cuvelai

Introduction

Until independence in 1989 the central-northern parts of Namibia formed the ethnic homeland "Ovamboland". The South African Apartheid government tried to isolate the area and its inhabitants as much as possible. This was not only a continuation of historical development[1] and in line with the general ideology of Apartheid, but also related to the war that South Africa fought in the area against SWAPO liberation forces. Namibia's independence opened up Ovambo society in a much more sudden fashion than in the case of other Namibian peoples.

While much of the traditional social and economic systems of the Ovambo remained untouched under Apartheid (Kreike, 1995), traditional forms of water supply were complemented by a huge pipeline network in order to ensure the supply of the new urban centres which were established mainly for administrative and military purposes. After independence, the extension of this network to improve water supply for the rural population became a major priority (Forrest, 2001). At the end of 1990, institutional reforms under the paradigm of IWRM complemented the infrastructure developments, aiming at the inclusion of local populations in the financing and management of the water supply. New institutions were set-up to co-exist with traditional forms of water management leading to institutional pluralism. In this regard, the water sector reflects the general transition of the Ovambo from a traditional society into a modern capitalistic one.

1 Already under German colonial rule a distinction was made between the so-called Polizeizone (police zone) and the Ovambo territory, which remained largely untouched by direct colonial rule (Tötemeyer, 1978).

This paper provides an analytical framework for analysing this transition.[2] It especially draws on the interaction between natural conditions, infrastructure and institutions which should allow for a more comprehensive understanding of these complex processes. It offers theoretical reflections about adequate mechanisms for the analysis of governance within coupled social-ecological systems. In section one, the field of study is outlined and is analysed as a coupled social-ecological system. In addition, the concept of resilience is introduced as a heuristic for understanding systems' ability to cope with external change. Section two presents the concept of social-ecological regulation. It serves as a meta-model for explaining change in human relations. The roles actors play in this model are outlined and how certain governance features influence actors' social learning processes as a result of social-ecological regulation.

Water Management within a coupled social-ecological system (SES)

Natural conditions and water supply

The study area is the Cuvelai basin in north-central Namibia. Mean annual rainfall is between 350 and 550 mm and occurs mainly between November and April (Mendelsohn et al., 2000: 9pp). The transboundary Cuvelai is a system of ephemeral rivers and small watercourses originating in southern Angola and flowing southward into the Etosha pan in Namibia. The high seasonality of rainfall together with the flat topography provokes recurrent alternations between floods and droughts in central northern Namibia.

Water is supplied from three major sources. Firstly, during the rainy season, the small depression fills with rain and/or surface water fed by floods from southern Angola. Depending on precipitation levels, these

2 The paper was developed as part of a Ph.D. thesis embedded in the CuveWaters research project <www.cuvewaters.net>.

incoming waters may lead to severe flooding. Due to the inter- and intra-annual rainfall variations, this water source is very unpredictable. Furthermore, high potential evaporation rates and above ground inflow of pollutants lead to problematic water qualities.

Secondly, area inhabitants use groundwater for domestic and agricultural purposes. In much of the Cuvelai basin, however, groundwater is saline (locally with TDS values exceeding 30 g/l) and is only fit for some specific uses. Therefore, many inhabitants take their water from relatively small, unconfined freshwater lenses which can be accessed by simple techniques like hand-dug wells. This technique reduces the potential danger of low water quality which might result from uncontrolled waste disposal, inadequate sanitation systems and other contaminants.

Thirdly, inhabitants of the Cuvelai basin receive water through a carrier from the Cunene river which marks the northwestern border between Namibia and Angola. The intake structure is located on Angolan territory at the Calueque reservoir. The water is directed through an open canal of about 160 km and finally distributed through a pipeline grid of about 2000 km in north-central Namibia (Kluge et al., 2006). Although this water is of much better quality than local groundwater, many villagers prefer to use latter, as it is cheaper. This supply network, however, represents a high level of dependence on the Kunene and the political and economic developments prevailing in Angola. For example, dry weather run-off in the face of increased energy production from hydroelectric power stations and the setting up and/or extension of irrigation farming could easily lead to supply bottlenecks.

Human relations with nature

The area under study is the most densely populated region of the country with about 42% of the total population living in 15% of the total area. Most people rely on smallholder agriculture including stockbreeding. Additionally, people have developed sophisticated agricultural patterns, adapted to the variable water availability. During the wet season, abundant surface water means that fishing is an additional source of income and dietary protein. Water – and its availability at the right time in the right quality – can be considered as the crucial determinant for the maintenance of local livelihood strategies. This is reflected in the various traditional rules, customs and institutions for water use and savings (Niemann, 2000).

Scientists predict that Namibia will be heavily affected by changing precipitation patterns due to global climate change. For northern Namibia (and southern Angola), a shortened and intensified rainy season with a precipitation increase between December and February and a subsequent decrease in March is predicted (Beyer, 2001). No major changes are expected is their geographical distribution (IPCC, 2007). Population growth, estimated to be about 2.6% (MoHSS et al., 2005: 7), together with rising water demand in northern Namibia and southern Angola are likely to lead to a decrease in water availability per capita; thus, the management of scarce resources will become more challenging in the future.

Social-ecological systems and resilience

This strong and direct dependence of people on a particular natural resource is the main reason why we have chosen to analyse this case as a social-ecological system.[3] Becker et al. (2006: 492) define a social ecological system as an "*analytical category, which allows for the formalized description and modelling of societal relations to nature.*" Social Ecology as an anthropocentric and multidisciplinary approach to practical science puts humans and their basic needs at the centre of interest. The key point for the analysis of SES is therefore the sustainable provision of ecosystemic services and the maintenance of livelihoods.

The capacity of SES to ensure essential services to people under changing external conditions – hence, to maintain its identity as a coupled SES – is described by the term resilience. People within the system (might) have the ability to regulate their relations with nature and therefore influence the adaptability of the SES and to keep it within a certain range of development that ensures the sustainable provision of ecological services

3 The underlying theory of Social Ecology is not at all limited to such obvious cases, but is, rather, a general theory of human-nature (Becker et al., 2006).

to inhabitants. This provision is the category that defines the identity of the analytical construct SES. A well-functioning SES permanently transforms itself in order to adapt to changes in external conditions. Its identity is based on the procedures which ensure this transformation. Its continuity is the permanent organisation of change.

Governance in social-ecological systems

Social-ecological regulation

External changes that influence SES might threaten traditional relations between humans and nature within the system and are therefore likely to be perceived as problems to actors.[4] Social Ecology emphasizes uncertainty in human-nature relations as an important factor for human activities. Management of natural resources under (changing) conditions of uncertainty requires what was handily put by (Berkes et al., 2003: 9) "*Be prepared for the unexpected.*" Uncertainty is a product of the complexity of functions and composites, which is a key attribute of systems. It is further assumed that human interventions in the system are often not predictable in their outcomes (Ratter, 2000: 44). As a result, every policy decision is taken under conditions of incomplete information about its possible consequences. Therefore, the solution to a particular problem cannot be more than an attempt requiring specific forms of feedback about its success. Dörner (1992) has pointed out that humans perceive the future usually as a linear continuation of the present; exponential developments are generally underestimated. It is therefore crucial how and whether feedback is channelled and how people deal with this information.

4 Social Ecology makes an important distinction between problematic situation, as an objective discrepancy between people's needs and abilities to act, and a problem, which is the subjective perception of the problem situation (Becker et al., 2006).

Feedback loops relate to the monitoring, detecting and solving of problems, i.e. of situations that were perceived as problematic by actors. In negative feedback loops, a problem is identified, approached and judged to be sufficiently, or not, resolved. If insufficiently resolved, a new approach to the problem has to be elaborated. If the resolution is found sufficient, problems of a second order derived from the solution might come into focus. The system thus transforms itself further through the permanent organisation of change. In social ecological thinking, a series of successful regulations, i.e. that which maintains the SES in a corridor of sustainability, is referred to as social-ecological transformation.

Box 1: Categories of power resources

	Remarks
exit costs	realistic exit options enhance the bargaining power (i.e. reduce the necessity for compromises)
risk behaviour	risk-seeking actors (who are better endowed with financial or other resources) can more easily
time preference	threaten commitments
credible commitment	patience wins
sanction power	closely related to social capital
organizability of a group	can oblige others to honour commitments better organized actors have more bargaining power than isolated actors
ideology	reduces bargaining power as it constrains the available alternatives
information and knowledge	information about changes and the knowledge to interpret and use the information is key in periods of transition
positional power	belonging to an established network strengthens an actor
transaction costs	costs to analyse what is at stake differ (relatively) among actors

Source: according to Knight (1996)

Actors in SES

Within this systemic model of how to regulate the permanent change in human-nature relations, actors engage in multiple interactions with each other as well as with natural resources. Such interactions occur on a lower level but relate to the overall systemic processes. Actors structure natural processes by building infrastructure and setting rules for their maintenance and by allocating the benefits generated. At the same time, nature constrains human behaviour through variables like soil carrying capacity or water availability. The analysis of SES has to identify the major natural and human-made components of the system and to describe their main features.

Actor-specific attributes, depending on their position, interests and power, determine actors' problem perceptions and hence their strategies in decision-making processes. Ostrom (2005) has pointed out that in a given problem situation actors do not generally behave uniformly according to the ideal types of social science: the self-interest driven egoist and the cooperative actor engaged in collective-choice agreements. A difficult interplay between structures and individual attributes is in fact found (ibid: 137). The identification of relevant actors and their specific attributes is therefore a crucial step in the analysis. (Knight, 1996), who argues for a rather conflictive view of institutional change, attributed the term "power" to actors' endowment with particular resources (*power resources*, see box 1). On the basis of these power resources, actors engage in a broad range of possible interactions including information-sharing among users of natural resources, deliberation processes, user conflicts, investment activities and lobbying activities (Ostrom, 2007: 15183).

We suppose that the endowment of key actors with power resources has a high explanatory value for their strategy in decision-making processes, leading to social-ecological regulation at the systemic level.

Contextual factors

Apart from actor-specific features, McCay (2002) has emphasised contextual factors as relevant for interactions in SES. She argues for a perspective on institutions that "includes patterned behaviour as well as rules and that locates institutions (=the rules of social interactions) as major features of the cultural, cognitive and ecological realms within which acting and decision-making individuals and social groups are embedded" (ibid: 362). With reference to anthropological theories, she emphasises "deep-rooted cultural biases", that pre-structure the behaviour of actors like egalitarianism, individualism, hierarchy and fatalisms. Enserink et al al. (2007) underpin this view with case studies from European river basins, where they demonstrate that cultural factors intensively shape the way people participate in decision-making. Factors like power distance, individualism, masculinity of society, uncertainty avoidance and long-term orientation co-determine peoples' willingness to engage in political processes.

Contextual factors can also be related to transition processes in the broader environment of the SES. This might be of particular relevance to the case study, as the institutional set-up of water management has changed dramatically since independence (Niemann, 2000: 129; Werner, 2007: 7; Kluge et al., 2007: 3). In such phases, formal institutions change very quickly a fact which can lead to a "lacking behind" of informal rules as they change only very slowly (North, 1990: 45) or not at all (in the respective period of time) (Saleth, 2004: 72). This can result in a difference between authoritative prescriptions and traditional behaviour, and the establishment of a grey area, which opens rooms for the redistribution of power and wealth. But informal institutions are also strongly influenced by the history of peoples' interactions, they are "carriers of history", reproducing themselves beyond the time of usefulness (ibid: 27). This survival of former rules or artefacts of rules, influences and partially directs the change of formal institutions. The phenomenon known as path dependency describes the fact that a certain trajectory of change is feasible according to the history of institutions. Revolutionary approaches that do not consider this history have little chance of becoming effectively implemented.

Furthermore, research on transition countries in former socialist countries has shown that transition periods provide specific conditions for powerful actors to use their resources (Theesfeld, 2005: 58). A common feeling of uncertainty can lead to the deterioration of social capital which is generally seen as facilitating the establishment and enforcement of institutions.

Governance for social learning

Due to multiple interactions, actors permanently re-frame their worldview. This re-framing is the basis for social learning, which has been defined as the substitution of one dominant institution by another (Tàbara et al., 2007). Actor-specific attributes, contextual factors and natural as well as infrastructural conditions sum up to a particular governance[5] framework in which social learning processes occur (Pahl-Wost et al., 2007). It is assumed that certain forms of governance structures foster social learning while others hamper it.

Polycentricity

Polycentric and multi-layered governance models with independent decision-making structures have been identified as laboratories for new policy approaches, (e.g.: Lebel et al., 2006; Moberg et al., 2005). It is argued that "an organizational structure with multiple, relatively independent centres creates opportunities for locally appropriate institutions to evolve by tightening monitoring and feedback loops and by enhancing associated institutional incentives" (Lebel et al., 2006). Polycentric and multi-layered structures also appear to be a logical translation of the multi-layered structures which are found in SES and therefore better able to reflect the properties of complex resources and to solve problems of fit (Scheumann et al., 2008). Additionally, Ostrom (2005: 236) stated that the "scientific management of natural resources" considered resources as homogeneous and large systems, which are linked over huge geographical scales. While this might be scientifically correct, it leads to a bias towards centralized

5 Rogers and Hall (2003: 4) as follows: "The concept of governance of course encompasses laws, regulations, and institutions but it also relates to government policies and actions, to domestic activities, and to networks of influence, including international market forces, the private sector and civil society. These in turn are affected by the political systems within which they function."

and uniform institutional structures in resource management, whether state-run or market-driven.

One important practical expression of polycentric governance is the establishment of so-called shadow networks (Olsson et al., 2006), often made up of actors from very different backgrounds, and which result from a growing perceptional convergence among actors (Dinar et al., 2004: 79). Their development is often triggered by crises (Pahl-Wostl et al., 2007). The main function of shadow networks is to allow for the exchange of different forms of knowledge about the problem and to encourage a common framing of reality. Successful changes in natural resource management are often connected to networks of scientists, ecologists and different practitioners.[6] Through this connection of diverse stakeholders and different types of knowledge, shadow networks have the potential to mediate between formal and informal institutions, i.e. between established views of human-nature interaction and newly evolved mental models. Pahl-Wostl et al. (2007) describe the process of institutionalization that can make informal (shadow) networks even "de facto a permanent part of the governance structure."

Participation

Polycentric governance structures should also facilitate the participation of a wide range of stakeholders which is supposed to be a precondition for social learning (Lebel et al., 2006). The goal is to ensure the inclusion of knowledge, information and different management approaches from stakeholders in the political process. According to the basic assumption that no stakeholder has sufficient knowledge to solve problems (Pahl-Wostl et al., 2007), it is very important to empower marginalized stakeholders to take part in these discussions. Additionally, it is hoped that people will come to trust authorities more, when they are actively involved in decision-making. Such trust in structures and decisions is important as it can increase the

6 The existence and importance of shadow networks for change processes has been analysed, for instance, in case studies from the Hungarian Tisza river basin (Sendzimir et al., 2008), the Swedish Kristianstad Vattenrike wetland (Gunderson et al., 2006) and the US American Everglades in Florida (ibid).

acceptance for adaptation measures. Therefore, the analyses of regulation cycles must require quality actor participation in decision-making.

Governance for social learning needs to accept water management as a highly political task that is too important to leave to a closed circle of experts (Adams, 1992: 36). In theory, the inclusion of different approaches should be guaranteed through open political discussions and the inclusion of as many concerned stakeholders as possible. Such forms of inclusive governance should allow for a permanent framing and re-framing of reality and facilitate the integration of different types of knowledge and information, especially 'scientific' knowledge and tacit knowledge (Scott, 1998: 313). It is obvious that this approach is demanding not only in political terms, but also with regard to the information needed. Gathering, processing and managing information about the SES becomes a political task as this is the basis for informed decisions. Actors need to understand cause-reaction chains and the interdependencies that structure their environment and influence their livelihood.

Furthermore participation is not a one-way street. On the one hand, stakeholders contribute their opinions and information, while on the other they receive information from the authorities about external influences and changes that that they were previously unaware of. Hence participation also offers the opportunity to raise awareness and build sensitivity for change. This requires a certain level of accountability from authorities in relation to the people. Decision-makers must justify their decisions and political processes must take place transparently. Beside this horizontal accountability, vertical accountability is equally necessary, me aning between experts (scientists/professionals) and non-experts. When different types of knowledge have to be included in a process, there is no place for superiority of one type above others. Participatory governance structures must build a framework in which this discussion and balance of knowledge can take place (Steyaert et al., 2007).

Communication is key to participatory governance. The relevant knowledge about the SES is distributed among actors and social groups depending on what their main occupations are (Pahl-Wostl et al., 2007). Crona et al. (2006) have shown with the example of Kenyan fishermen that communication ties follow mainly these occupational structures. They identified stakeholders who had ties to different occupational groups and thus served as hubs in the relational network. The centrality of a communication network, i.e. the existence of hubs that have more linkages to other elements of the networks than others, is therefore an important attribute of the whole SES. It is important to note that the existence of hubs is not *per se* good (or bad). When distributing information quickly in the network, they foster social learning and might enhance the ability of the SES to react quickly to changes (Longstaff et al., 2008). But there are also examples of highly centralized SES in which diseases spread quickly due to the existence of major hubs (Crona et al., 2006). Broadly speaking, the problem perception of people differs according to the information they have and according to the density of communication they maintain with members of other occupational groups. Therefore, communication-enabling hubs are an important factor for social learning.

Another important aspect of participatory governance is leadership, which has been identified in case studies as crucial for successful social learning (Mostert et al., 2007). While different problem perceptions are discussed, it is necessary for "political entrepreneurs" to take responsibility and bring forward the formulation of a perceptional convergence (Saleth et al., 2004). Leadership is critical in social learning, because it helps balance differing opinions in multi-stakeholder processes. Communication, conflict management, initiation of partnerships and the development of visions are therefore crucial tasks for leadership (Mostert et al., 2007). Leadership is also required to bring stakeholders together into shadow networks, so as to formulate interests and strategies. It also has the function of bridging gaps between shadow networks and formal institutions.

Conclusion and further questions

In this paper, we have analysed governance in SES by examining actor-specific and contextual factors as well as the particular governance structures that shape the possibilities for social learning.

We have pointed out that polycentric and participatory governance is supposed to benefit the establishment of feedback cycles and social learning of actors. In theory, this should allow for successful social-ecological regulation. We can assume that we will not be able to identify such ideal types in reality. But it might be possible to identify traces of them and relate them to successful regulation processes. An empirical analysis based on these theoretical reflections might therefore allow for a clearer identification of key drivers or obstacles to successful social-ecological regulation.

Therefore, these theoretical reflections suggest further research questions such as:

- To what extent does the given governance system differ from the ideal type of polycentric and participatory governance?
- What are the main (actor-specific or contextual) factors hindering social learning?

References

Adams, W. M. (1992). *Wasting the rain. Rivers, People and Planning in Africa.* Minneapolis / London: University of Minnesota Press.

Becker, Egon and Jahn, Thomas (eds) (2006). *Soziale Ökologie. Grundzüge einer Wissenschaft von den gesellschaftlichen Naturverhältnissen.* Frankfurt/New York: Campus Verlag.

Berkes, Fikrit; Colding, Johan, and Folke Carl (eds) (2003). *Navigating Social-Ecological Systems. Building Resilience for Complexity and Change.* Cambridge: Cambridge Universtity Press.

Beyer, Ulrike (2001). *Regionale Niederschlagsänderungen in Namibia bei anthropogen verstärktem Treibhauseffekt – Abschätzungen mit statistischem Downscaling.* Würzburg.

Crona, B. and Bodin, Ö. (2006). What you know is who you know? Communication patterns among resource users as a prerequisite for co-management. *Ecology and Society* 12(2).

Dörner, Dietrich (1992). *Die Logik des Mißlingens. Strategisches Denken in komplexen Situationen.* Reinbeck bei Hamburg: Rowohlt.

Enserink, Bert, Patel, Mita, Kranz, Nicole, and Josefine Maestu (2007). Cultural factors as co-determinants of participation in river basin management. *Ecolgy and Society* 12(2).

Forrest, Joshua (2001). Water policy and environmental sustainability: The case of post-colonial Namibia. *Public Administration and Development* 21 (5): 393–400.

Gunderson, Lance, Carpenter, Steve, Folke, Carl, Olsson, Per, and Petersen, Gary (2006). Water RATs (Resilience, Adaptability, and Transformability) in Lake and Wetland Social-Ecological Systems. *Ecolgy and Society* 11(1).

IPCC (2007). *Climate Change 2007: Synthesis Report. Contribution of Working Groups I, II and III to the Fourth Assessment Report of the Intergovernmental Panel on Climate Change* [Core Writing Team, Pachauri, R.K and Reisinger, A. (eds.)]. Geneva: IPCC.

Kluge, Thomas, Liehr, Stefan, Lux, Alexandra, and Moser, Petra (2007). *Integrated Water Resources Management in Central Northern Namibia.* Frankfurt: ISOE.

Kluge, Thomas, Liehr, Stefan, Lux, Alexandra, Moser, Petra, Niemann, Steffen, Umlauf, Nicole, and Urban, Wilhelm (2006). IWRM Concept for the Cuvelai Basin in Northern Namibia. *Physics and Chemistry of the Earth* 33: 48–55.

Knight, Jack (1996). *Institutions and social conflict.* Cambridge: Cambridge University Press.

Kreike, Emmanuel (1995). *The ovambo agro-silvipastoral system: Traditional land use and indigenous natural resource management in northcentral namibia.* Windhoek: Republic of Namibia, Directorate of Forestry, Ministry of Environment and Tourism.

Kundzewicz, Zbigniew (2007). Climate-related water phenomena: which regions are most affected? *EZ trifft Wissenschaft: Anpassung an den Klimawandel.* Eschborn: GTZ.

Lebel, L., Anderies, et al. (2006). Governance and the capacity to manage resilience in regional social-ecological systems. *Ecology and Society* 11 (1).

Longstaff, P. H. and Yang, S. (2008). Communication management and trust: Their role in building resilience to "surprises" such as natural disasters, pandemic flu, and terrorism. *Ecology and Society* 13 (1).

McCay, Bonny (2002). Emergence of Institutions for the Commons: Contexts, Situations and Events. In E. Ostrom, T. Dietz, N. Dolsak et al (eds), *The Drama of the Commons,* p. 361–402. Washington, DC.: National Academy Press.

Mendelsohn, J., S. El-Obeid, et al. (2000). *A profile of north-central Namibia.* Windhoek: Gamsberg Macmillan Publishers.

Moberg, Frederik, Galaz, Viktor (2005). *Resilience: Going from Conventional to Adaptive Freshwater Management for Human and Ecosystem Compatibility.* Stockholm: Swedish Water House.

MoHSS and WHO (2005). *Equity in Health Care in Namibia. Towards needs-based allocation formula.* Equinet Discussion Paper No. 26. Windhoek.

Mostert, E., C. Pahl-Wostl, et al. (2007). Social learning in European river-basin management: barriers and fostering mechanisms from 10 rivers. *Ecology and Society* 12 (1).

Niemann, Steffen, (2000). *Wasserversorgung und Wasserverwendung in Namibia. Nutzungstraditionen als Grundlage eines nachhaltigen Ressourcenverbrauchs im ehemaligen Ovamboland.* Hamburg: Institut für Afrika-Kunde.

North, Douglas (1990), *Institutions, Institutional Change and Economic Performance.* Cambridge: Cambridge University Press.

Olsson, Per, Gunderson, Lance, et al. (2006). Shooting the rapids: Navigating transition to adaptive governance of social-ecological systems. *Ecology and Society* 11 (1).

Ostrom, Elinor (2005), *Understanding Institutional Diversity.* Princeton and Oxford: Princeton University Press.

Ostrom, Elinor (2007). A diagnostic approach for going beyond panaceas, *PNAS* 104 (39): 15181–15187.

Pahl-Wost, Claudia, M. Craps, et al. (2007). Social Learning and Water Resources Management. *Ecology and Society* 12 (2).

Pahl-Wostl, C., J. Sendzimir, et al. (2007). Managing Change toward Adaptive Water Management through Social Learning. *Ecolgy and Society* 12 (2).

Ratter, Beate (2000). *Natur, Kultur und Komplexität. Adaptives Umweltmanagement am Niagara Escarpment in Ontario, Kanada.* Berlin, et al.: Springer.

Saleth, Maria and Dinar, Ariel (2004). *The Institutional Economics of Water. A Cross-Country Analysis of Institutions and Performance.* Washington: World Bank.

Scheumann, Waltina and Herrfahrdt-Pähle, Elke (eds) (2008). *Conceptualizing Cooperation for Africa's Transboundary Aquifer Systems,* Studies 32. Bonn: Deutsches Institut für Entwicklungspolitik (DIE).

Scott, James (1998). *Seeing like a state: How certain schemes to improve the human condition have failed.* New Haven: Yale University Press.

Sendzimir, J., P. Magnuszewski, et al. (2008). Assessing the Resilience of a River Management Regime: Informal Learning in a Shadow Network in the Tisza River Basin. *Ecolgy and Society* 13 (1).

Steyaert, Patrick and Ollivier, Gerard (2007). The European Water Framework Directive: how ecological assumptions frame technical and social change. *Ecology and Society* 12 (1).

Tàbara, J. D. and Pahl-Wostl, C. (2007). Sustainability learning in natural resource use and management. *Ecology and Society* 12 (2).

Theesfeld, Insa (2005). *A Common Pool Resource in Transition. Determinants of Institutional Change for Bulgaria's Postsocialist Irrigation Sector.* Aachen: Shaker Verlag.

Tötemeyer, Gerhart (1978). *Namibia old and new. Traditional and modern leaders in Ovamboland.* London: C. Hurst & Co.

von Braunmühl, Claudia (2005). *Water Governance – Partizipation in der Wasserversorgung.* Networks Papers 18. Berlin: DIFU.

Werner, Werner (2007). *Integrated Land and Water Management: Policy and Institutional Issues, CuveWaters Papers No. 1.* Frankfurt-M.: ISOE.

Signe Marie Cold-Ravnkilde*

Between a rock and a hard place: Decentralization and struggles for water in Mali

Abstract. Access to water is vital to local people's livelihood strategies in rural Africa, and the way in which water is controlled determines who and under which conditions people get access to water for domestic and productive purposes. This chapter presents a case study of a lake situated in the central part of Mali, which became permanent in 1991 possibly due to a "hydrological paradox". The lake is located in an arid environment characterized by local political struggles over scarce water resources amplified by decentralization in Mali. The presence of the lake significantly changed the opportunities for both the population living there and the municipalities striving for political influence. The case study reveals that local actors mobilize and counter-mobilize identities of belonging to create a winning coalition for control of the lake. This chapter presents neither a pessimistic nor an optimistic view of water in Africa. Rather I argue that conflicts over access and control of water resources are part of everyday local politics in Mali. Consequently, the study of conflict over access and control serves as an entry-point into comprehending the practices and processes of local governance from a rural point of view.
Keywords. Conflicts, access, control, decentralization, Mali

Introduction

In Mali's post-independence period natural resource management was characterized by central government interference, open access regimes,

* Danish Institute for International Studies, and International Development Studies at Roskilde University. Strandgade 56, 1401 Copenhagen, Denmark, smr@diis.dk.

and conflicts between government agencies and farmers over the management of common pool resources (Moorehead, 1991; Degnbol, 1999; Vedeld, 2001). From the beginning of the 1990s a wider process of democratization, envisaging that enhanced political participation at the local level might strengthen democracy, sustainable development and poverty reduction took place in Mali. International donors have often pointed to Mali as a successful case for implementing democracy in Africa mainly because democracy was claimed to be a demand from the population and a pragmatic political response to the Kel Tamasheq rebellion[1] in the northern regions (Gaasholt, 2009; Seely, 2001; Boone, 2003; Ouedraogo, 2003).

However, as in other places (Boone, 2003; Crooks and Manor, 1998), in Mali decentralization has not always produced the desired effects and the real contribution of decentralization has been questioned by many (Bocoum et al., 2003; Coulibaly and Hilhorst, 2004; Djiré, 2004; Oluwu and Wunsch, 2004). As part of the donor-driven decentralization reform in Mali, various administrative competences of the state, including those concerning water management, have been transferred to local governments[2] (Hetland, 2007; Nijenhuis, 2003; Zobel, 2004; Cissé et al., 2003). In the light of earlier centralist regimes, trends towards decentralization and democracy were seen as new and unrivalled opportunities for more legitimate local governance and local participation in natural resource management (Benjamin, 2008; Ribot, 2004; Vedeld, 2001). The process of decentralization has made the management of common property water resources to become more complicated (Bocoum et al., 2003; Kassibo, 2004). Furthermore scholars have shown how municipalities have become embedded in existing social relations and local politics (Kassibo, 2004; Hetland, 2007; Hanonou, 2009) and that the municipalities have only provided new forums for already existing powerful actors to benefit through clientelistic practices while the rural people's participation in development processes is still lacking

1 The Kel Tamasheq are also known as Tuaregs and make up six percent of Mali's population (Gaudio, 1988). I apply the politically correct term refer ing to people speaking the language of Tamasheq (see Hanonou, 2009).

2 Decentralized units in Mali are called collectivités territoriales and are defined on the levels of region, district, municipality and the district of Bamako (Hetland, 2007:12). At the village level, the village chief or the tribal chief constitute extensions of the state administration (Hetland, 2007; Gaasholt, 2009).

(Bakayoko, 2007; Djiré, 2004; Hetland, 2007; Boone, 2003). The crucial issue here is that decentralization and the creation of municipalities are changing the institutions of public authority engaged in natural resource management, and, particularly, water management.

This chapter shows how water conflicts are configured within the institutional changes provided by decentralization in Mali. Here, neighboring municipalities share valuable water resources and competition for water is high. The chapter argues that conflicts over water are part and parcel of local politics in the Sahel and can only be fully understood within this framework. The conflicts over *Lake Aman*[3] are prevalent both in competing claims over which municipality has the legitimate authority to control the lake, and in how people gain access to the lake. The nature of this dispute indicates that in Mali, like in many post-colonial countries, *struggle over property is as much about the scope and constitution of authority as about access to resources* (Lund and Sikor, 2009: 2). The water resource in question is Lake Aman which became a permanent lake in 1991 due to climatic changes. It is situated in an arid environment where water is scarce and is therefore crucial for people's livelihood strategies.

The chapter is organized as follows: first the geographical, socio-political and historical background is provided, which explains why water is an important political issue in this area. Next, conflicts over Lake Aman are presented in two situations. The first illustrates how municipalities contest each other's boundaries in order to gain ownership of the lake; the other shows the difficulties of seasonal migrating cattle herders in accessing water and pastures for their animals at Lake Aman.

The empirical data presented are based on fieldwork from a current Ph.D. research project on water conflicts in central Mali (Douentza district). The field data were gathered in October 2007, February 2008, and from October to March 2009 in Aman village as well as in Akal and Koyra. The methods used were mainly qualitative. In addition to interviews, statistical data were also used within the framework of the Ph.D. research.

3 The name of the lake and the village has been changed. Aman means "water" in Tamasheq. The two rural municipalities discussed are referred to as Koyra which means "village" in Songhay (Mauxion, 2008) and Akal which is often translated as "earth", "land" or "country" in Tamasheq (Berge, 2001). I also gave pseudonyms to people cited.

Who owns Aman?

Lake Aman is situated in a valley on the border between two neighboring rural municipalities in central eastern Mali. *Koyra* municipal headtown lies 30 km from Lake Aman in the Douentza district within the Mopti region. *Akal* municipal headtown lies 55 km north of the lake in the district Gourma Rharousse within the Tombouctou region.

The region has a typical Sahelian-Saharan climate, with one rainy season with an annual average rainfall of 400mm characterized by inter- and intra annual fluctuations (Diarra, 2008).[4] Surface water resources dry out in the dry season where water is only available at dispersed wells. However, possibly due to the *Sahelian Hydrological Paradox,* Lake Aman has become permanent after a year of heavy rains in 1991.[5] The paradox explains that due to drought-caused loss in vegetation and increased run off, ground water raises, while small ponds occur in low-lying areas (l'Hote et al., 2003; Leblanc et al, 2008; Benjaminsen 1993; Mougin et al, 2009).

In 1998, a village was founded next to the lake. With the arrival of the municipalities in 1999, Lake Aman has become a point of conflict between the municipalities of *Koyra* and *Akal* who each want to claim the revenues of the new opportunities provided by the lake such as fishing, planting of *bourgou*, gardening and local trade.

The national road runs through Lake Aman. On the shores of the lake the people cultivate gardens and small shops are emerging next to the road. Aman is thus considered as a village although it has not yet received official recognition. The presence of the lake and pastures of Aman valley have attracted people from many different places. Because Aman used to be a pastoral site where people did not settle permanently, the origin of

4 The 1950s were unusually good years in terms of rainfall. The years 1973 and 1984 were extremely dry years and had a strong negative impact on the area's ecology and population, which suffered from insufficient pastures and crop failure (Pedersen, 1995). The years of 1988–1991 have been relatively good years in terms of rainfall (Benjaminsen, 1993; Diarra, 2008).

5 The emergence of new ponds in the Gourma is not a unique pheonomenon but the causes are not yet fully understood. See Mougin et al, 2009 for a thorough discussion of the occurrence of Aman and the dynamic and changing character of the Sahelian climate.

Aman is difficult to define and thus narratives of belonging have become strategic weapon in the conflict over the valuable territory.

To categorize people ethnically in this area is a challenge due to the complex ethnic composition and different periods of settlement during the big Sahelian drought of the 1970s and 1980s. Furthermore many refugees have settled in the aftermath of the Kel Tamasheq rebellion in Mali towards the end of the 1990s.[6] Although none of the tribes in Aman have been rebellious themselves, the conflict has strongly affected their lives.

Each tribe has its own relationship of belonging according to historical, social, and political conditions: some tribes administratively belong to the municipality of Koyra while others belong to tribes under the administration of Akal. This complexity reveals the difficulties in defining which community to belong to in a pastoral setting where people due to a number of external chocks and unstable climatic conditions are used to being on the move and in search for "greener pastures" (de Bruijn and van Dijk, 2003).

In the history of the region, there have been various conflicts and alliances between the different ethnic groups over access to scarce natural resources. The recent conflicts over Lake Aman dealt with here started in 2000, right after the establishment of the municipalities. The boundaries between the municipalities had not yet been defined and, because Lake Aman is situated on the border between two neighboring municipalities, the mayors started writing letters to each other, arguing that the lake is situated within their municipal territory and that consequently the land, the population living there, and the economic revenues from the activities belong to them. Furthermore, the grazing areas and the vegetation cover in Aman is decreasing and, as we shall see, migrant Fulbe herders who are defined as "outsiders" are struggling to get access to Lake Aman, an access which has become difficult to obtain with the strengthening of community-based natural resource management. Before describing the conflicts, a presentation of the different parties involved is provided as

6 The rebellion is one in a series of Kel Tamasheq revolts in the colonial and post-colonial history in which the Kel Tamasheq claim that successive Malian governments have all largely ignored their interests, both in terms of access to development aid, economic development, and violations of human rights (Seely, 2001).

the conflicts can only be fully grasped within their particular social and historical setting.

Chiefs, settlement, and belonging in a post-conflict context

The population living in Aman consists of around 150 households who live there more or less permanently throughout the year. In the rainy, large herds are brought in from the Inner Niger Delta to graze at Aman but in some years the absence or limited availability of pastures forces the herders to go elsewhere.

Songhay farmers live in proximity to Fulbe and Kel Tamasheq groups that used to be nomadic herders but recently have become more sedentary, although herding is still their main activity.

In the colonial era the Songhay from Koyra collaborated with the French in the battle against the Kel Tamasehq from Akal. During this battle one of the pastoral tribes now living in Aman, the *Foulankriabé*,[7] fled from the Kel Tamasheq who used the Foulankriabé as artillery. Because of this treatment the Foulankriabé sought protection from the Songhay chieftaincy in Koyra (Marie, 1975) who dug a well and gave the Foulankriabé user rights to the pastures in Aman (Marie, 1975). The leader of Foulankriabé claims that his people are the first occupiers of land and legal tenants of the customary rights to the land in Aman.

> The first well was created by my father. At that time there were pastures but no water. They could be without water for 24 hours. It was in the colonial period. The Iderfanes (another Kel Tamasheq Tribe) say they are the first comers, but that is not correct. (leader of the Foulankriabé, Aman, 2009)

However, with decentralization, the municipalities have realized the potential of Aman and given the weak social position of the Foulankriabé, their influence has diminished, as they have been pushed away from their original position close to the lake.

7 The Foulankriabé are of Fulbe origin (Gallais, 1975) but speak Tamasheq.

The *Iderfanes* are Kel Tamasheq[8] and belong to the Akal municipality. Their chief also claims to be customary chief and that large parts of the land belong to him.

> I am born in Aman like my father and my grand father. My father was tribal chief since the colonial times when there was not even a house in Akal and you paid your taxes in Rharouse. Now people pay their taxes to me and I will go on camel back or by foot to collect tax. (Iderfanes chief, 2008)

Although living in dispersed areas the Iderfanes are one of the largest tribes in Akal and play a key role to the municipality among other things because of tax income. In 2000 an informal market for selling land emerged in Aman and the Iderfane chief also started to sell the land which triggered a dispute between the two municipalities over who had the authority to issue the land titles.

After the droughts of the 1980s, a western aid organization intervened in Akal to help, in particular, the pastoral population who had lost their livestock. In this context a third tribe, the *Imrad*,[9] and their tribal chief first settled in Aman with help from the western aid organization. The Imrad chief claims he was the first to build a house, which was washed away by the rains in 1991. With the Kel Tamasheq rebellion in Akal in 1992, the Imrad chief sought refugee in Burkina Faso where access to development aid enriched him and strengthened his position as chief.[10] The Imrad came back in 1996, but have been harassed and tortured by the Malian military, which explained their often hostile appearance in the village. The Imrad asked to be a separate tribe to gain power in Aman and have monopolized one part of the Lake. The chief of Imrad also claims to be chief of Aman and controls several activities in the village.

The Imrads are often in dispute with the Fulbe from the Inner Delta who come to graze in the rainy season in Aman.

8 Local people often make use of the red/black contrast to define ethnic and moral differences (see Hanonou, 2009: 176). The Iderfan former slaves of the Kel Tamasheq were freed in the past. However, because people know they were slaves, their origin is still associated with their identity (Bernus 1981: 75).

9 Imrad is a pseudonym which means a "free man" in Tamasheq and refers to the high social ranking of this tribe (Gallais, 1975).

10 In 2000 the Norwegian Church Aid withdrew completely from financing the local NGOs who had financially supported political parties in Akal.

There is also a Fulbe tribe from Burkina Faso who settled more permanently in Aman after the drought in the 1980s. Their main activities are live stock herding although some also practice rain fed agriculture in the rainy season. The Fulbe are often regarded as rich because they possess a large number of animals. However, politically, their influence in the dispute about Lake Aman is limited, and they often prefer to stay out of trouble with the other groups.

The *Songhay* groups in Aman administratively belong to Koyra. Historically, they have engaged in agriculture and adopt a sedentary livelihood strategy. During the rebellion in 1992, their leader fled from a neighboring pastoral site where the Western aid organization had engaged him as nurse. He settled around Lake Aman in 1998 and also presents himself as chief of Aman. He tells me that he constructed the first houses, the mosque and created a village association (personal communication). Furthermore, a few lucky individuals from his group have started to practice horticulture in enclosed gardens on the shores of the lake. The Songhay leader also started to sell land titles in "his village", which the municipality of Akal prevented. Other Songhay families have engaged in fishing and have made agreements with the municipality of Koyra concerning the exploitation of the fish stocks. The different livelihood strategies employed in Aman create tension among the inhabitants, and cohabitation is a challenge in Aman which has only been further complicated by the involvement of administrative authorities.

Main water uses and water infrastructure employed

The population use water for drinking, cooking and cleaning as well as economic activities such as livestock, horticulture and fishing. Furthermore with help from a government-funded project, a sea grown fodder plant, the *bourgou*, has been planted inside the lake to provide fodder in the dry season.[11]

11 The *bourgou* is one of the most prized resources in the, see Cotula, 2006.

A couple of wells have been created in Aman with little success. The first was built in 1936 during the colonial occupation. Although no longer functioning, this facility still serves as a strategic reference point in the conflicts over Aman.

Since Lake Aman became permanent, most households have used the lake as their primary water source. The water is of poor quality and some households use filters and local products to purify water. In 2008 the Aman received a hand pump with the help of a French research team.[12] Despite the pump, some people still prefer water from the Lake, since it is easier to cool down in the hot season and many people think it tastes better and that access is easy (Household survey, 2009).

In terms of issues of water shortage, the three customary chiefs of the Iderfanes, the Imrad, and the Songhay are informed and consulted. Each of them represents their community and manages different activities. In general, water from the Lake is regarded as the property of the Gods and only large herds are limited in accessing the lake due to fear of contamination from sick animals and over-grazing. However, access rights to highly valuable plots on the shores of the lake and access rights to fish are strictly regulated.

To sum up: there are multiple stakeholders with various interests engaged in the conflicts over Lake Aman. The production patterns adopted by the various groups are significant for water resource access and control. The permanent inhabitants use the lake for domestic and productive uses. The temporary inhabitants are the Fulbe from the Delta who seek access to water and pastures in the rainy season. The Fulbe practice seasonal mobility between the wetlands of the Inner Delta of the Niger River and the dry lands to the east.

The customary chiefs want to establish themselves as chiefs of Aman to control the land and the water-related activities. The municipalities of Koyra and Akal both want taxes, votes, and revenue from the economic activities created by the lake. With these different actors and their interests in mind, we now turn to the conflictive situations.

12 It was impossible to determine the financing and date of drilling of this pump (personal information, 2008).

Contested territorial boundaries

The dispute over which of the two municipalities, Koyra or Akal, the lake belongs to started in 2000 when both mayors started to write official letters to each other and to their superior, the prefect,[13] claiming ownership of the lake.

The situation escalated when Mali's ministry of decentralization delegated a mission to set up an entrance gate on the national road to mark the entrance into the Tombouctou region. The gate served as a physical proof that Lake Aman would be situated within the municipality of Akal (Tombouctou region). When a group of men from Koyra were informed that Lake Aman had been "given" to Akal they performed a protest march to reclaim the lake. The mayors of the two municipalities, their counselors and the state representative of the two districts gathered to discuss the issue. During the meeting more tension broke out and no agreement on the ownership of the lake was found and the entrance gate was abandoned. The minister of decentralization later proposed to divide the lake in two. This has not been considered a solution by the people living by the lake and the question has never been officially resolved, although Akal, due to continuous on site investments, is gaining territory.

The situation is complex because the contested claims over Aman occur in a context where new institutions of public authority are emerging and struggling to consolidate their power in the process of decentralization. Although based upon previous administrative boundaries of the cantons, in Mali the municipal boundaries were not defined, because the Malian state was afraid that the delimitation itself could further exacerbate local conflict. The blurred municipal boundaries in absence of statutory decision-making create an opportunity for the contesting municipalities to gain territory.

Lake Aman is a natural open water source of which a municipality cannot claim ownership since lakes are state property and of public use

13 The Prefect ensures that municipal proceedings follow government legislation and authorizes the release of central government funds for municipal development plans.

(République du Mali, 2002).[14] However the municipalities can claim the territory surrounding the Lake and thereby make legitimate claim to the benefits derived from the resources and opportunities the Lake brings. In the dispute over which of the municipalities Lake Aman belongs to, the Akal municipality claims that Aman is situated within the territory of Tombouctou according to a national map from 1959. In one of the many confidential letters from the mayor he writes:

> I respectfully complain to your authority regarding the actions taken by the Mayor of Koyra at Aman. Since the arrival of the municipalities, my counterpart has not ceased to provoke me in regard to the ownership of Aman. In regard to the instances of the State we have always politely asked him to consult the colonial map from 1959. In his stubbornness he personally goes to Aman to stock fish in the Lake. Something we did four years ago. After intimidating our agents and our represenatives at Aman he shows up in the fields, stocks fish in the Lake, and sets out a plot with a view to achieve a project. We protest against these acts which we consider null and void. The Lake, its content, its riverside residents belong to our municipality. (Letter from the Mayor of Akal to the prefect in Douentza, 2003)

The letter shows that the investment in fishing stocks that the Songhay have made in collaboration with the Koyra municipality is indirectly a way of claiming ownership and authority in Aman, which obviously provokes the mayor of Akal.

Koyra, on the other hand, claims that Aman belongs to them according to a pastoral convention from 1939. The convention defines the grazing limits for the Kel Tamasheq from Akal and the Foulankriabé. And because the Fouankriabé belongs to Koyra, Koyra claims the territory where Aman is situated today.

> I attract your attention to the inconveniences that would follow from the distribution of the land of Aman… it is necessary to remember: that the place depended on the ex-canton of Koyra that I personally controlled temporarily in 1952; that the well of the site is drilled under the auspices

14 In principle, water is regarded as a public resource and cannot be privatized. Article 3 deviates from this disposition provided that the privatization is carried out according to the law and in respect of the customary authorities (République du Mali, 2002: art 3; Cotula, 2006:13–15; NCG, 2007).

> of this same canton in 1936; that the official limit between the ex-canton of Koyra and the Kel Akal tribe is located well far to the north of Aman. (Former mayor of Koyra, 2000)

In the dispute a number of issues are crucial.

First, history is instrumentalized to legitimate ownership of Aman today. The mayor of Koyra strategically uses the situations of the colonial past to establish a fixed point for the present. When I ask the chief of the Iderfane about the official limit, which the mayor in Koyra refers to, he says:

> (the convention) is now cancelled. Aman belongs to Akal. There was even a signpost made of iron before independence. It is on the basis of this signpost that the colonials have put on the map that Aman belongs to Akal. (Iderfane Chief, Aman, 2009)

The different narratives show that boundary making is a strategic positioning which draws upon historical practices to create a repertoire of meaning defining who is in control in a changing environment (Li, 2000). However these points are by no means fixed, and continuously reinterpreted according to the purpose they serve.

Second, to strengthen the legitimacy of the municipalities, the ability to deliver public such as wells, fishing farms, vaccination parks and development projects become means to legitimize the ownership of Aman. The capacity to provide services and investments is part of the strategy to gain legitimacy as the public authority in the administrative flux of the present. For Akal the development activities carried out by a western aid organization became a claim in the dispute (Mayor of Akal, 2007). For Koyra the fish stocks were the only significant investment beside the well from 1936, which no longer functions. Akal has invested in a vaccination park and a community school which has de facto legitimized Akal's claim *vis-à-vis* Koyras reference to colonial history.

Third, having citizens in Aman becomes another way for the competing municipalities to obtain authority.[15] Both municipalities claim

15 The Songhay, the Fulbe and the Foulankriabé are enrolled in Koyra, while the Imrad and the Iderfanes are enrolled in Akal.

that the people living in Aman administratively belong to their municipality and that Aman belongs to them. Therefore, creating citizens is a way for the municipalities to win territory, votes, taxes, and water resources.

Struggles over access to water

For centuries Fulbe herders from the Inner Delta of the Niger River have come to graze their animals in Koyra during the rainy season while the Inner Delta is flooded. The seasonal movement of livestock between the floodplains and the dry land pastures is essential to the Fulbe herders' livelihood and herd mobility has shown to be a well-suited livelihood strategy for the harsh and unpredictable environmental conditions in the Sahel (Thébaud, 2000; Benjaminsen, 1997). The seasonal movements of animals require negotiations between the transhumant Fulbe herders and the residents over access to pastures and water points (Thébaud, 2000; Cotula, 2006; Cold-Ravnkilde, 2009; de Bruijn and van Dijk, 1995). However, the reciprocity between herders and sedentary populations has deteriorated many places in the Sahel as more farmland has come under cultivation at the price of pastoral corridors (de Bruijn and van Dijk, 1995; 2003; Beeler, 2006; Turner, 2004).

The second conflictive situation took place in Aman in October 2007 and illustrates the consequences for access to the lake for non-resident herders. The migrant Fulbe struggle to gain access to water and pastures in Aman. The situation started because a migrant Fulbe herder watered his cattle and crossed the recently planted *bourgou* plot belonging to the Imrad chief.

> We have our plots, which we want to exploit, but we don't have the means to buy the fence to close them off. There are always animals destroying the plots. Our resource is the lake, but we cannot use it because there are too many animals grazing around the lake. Sometimes more than 15,000 cattle come in the rainy season. They come with their sick animals, and they eat up all the pastures. (Chief of Imrad, 2008)

The herder continued to let his animals enter the water despite the threats from the Imrad chief, and one day, in the heat of the moment the herder drew a gun. The chief informed the gendarmes from Akal who arrested and fined the herder for carrying a weapon. However a solution for the transhumant herders had to be found and the problem was transferred to the sub-prefect (state representative at the municipalities) in Koyra who indicated a passage corridor for the cattle which did not cross the *bourgou* plots. The passage corridor was respected; but similar incidents occurred the next year, when a new group of herders arrived at Aman.

The situation illustrates how herders struggle to gain access to water for their animals at Aman and reveals that gaining access to the lake for "outsiders" has become more difficult after the settlement of the Imrad who have invested in *bourgou.*[16] In this context it is not so much water but grazing and fodder resources which are scarce; nevertheless in Sahel water and pastures are interrelated and access to pastures and grazing is often regulated by regulating access to water (Thébaud, 2000; Cotula, 2006; Cold-Ravnkilde, 2009).

Several factors are crucial here. The Imrad have strong ties to Akal and have been benefitting from the investment in the planting of *bourgou.* The community-based management of the *bourgou* has further limited the access of the migrant herders who are not citizens of Aman while the increased competition for resources has made belonging to the community an instrument for securing access to the valuable resources. The *bourgou* project had thus reinforced the position of the Imrad, and strengthened the ability of the chief to force herders to leave.

The ownership of the *bourgou* and the Imrad chief's attempt to limit access to Aman is however contested by the Fulbe. A herder who has come to Aman for the last 20 years explains the situation:

> When we first came to Aman, no one prevented us from watering our animals, because there was not even a village here. But for the past four years, the people in Aman have started to show us where we can water our animals. It is the Iderfane chief who shows us where we can go with our animals, but we have refused. They hit us and insult us, but we still demand

16 Similar situations have been seen other places in the Douentza district where migrant herders have been excluded from watering their animals.

> one law (e.g. for all). There is no difference between where they want us to go and where they refuse to let us water our animals. What bothers us is that we knew the lake before the Iderfanes were here in Aman. They came last. The Iderfanes will go to Akal. But we don't care whether it is Akal, Koyra or all over Mali, because it is still within the territory of Mali (Fulbe herder, Aman 2008)

The herder claims to have been in Aman before the Iderfane, in order to legitimate equal access to the lake as well as to delegitimize the actions of the Iderfanes. The herder simultaneously makes claim to his national citizenship and the equal law for all within the territory of Mali. This way he contests the privileged access that belonging to the community is claimed to provide.

Usually access to ponds, particularly in the rainy season is regarded as free to the herders, since surface water sources like ponds belong to the state and cannot be privatized (République du Mali art, 3; Thébaud, 2000; Cold-Ravnkilde, 2009). When surface waters dry up, pastoralists move to groundwater points, such as wells and boreholes to which access is negotiated and payment in cash or kinds is often required. Changing the patterns of not paying for accessing an "open" water resource might take time for the herders, and also renders their livelihood strategy of seasonal mobility costly. However, local people have often pointed to the inequality of circumstances between the fact that herders come from the Inner Delta,[17] where access to water and pastures are strictly regulated, and that locally they have been able to access water and pastures without payment.

What further weakens the Fulbe herders' position *vis-à-vis* the chiefs is that within the social organization of the Fulbe it is usually the young men who conduct cattle on long distance migrations. Sometimes the herds belong to rich businessmen from the main towns and the herder is hired to take care of the animals. As a result, the herders have little authority to negotiate with local chiefs for access to resources, and they are excluded from both decision-making processes and free access to the lake.

17 The management of the Delta constitutes a special case. See van Dijk, 1995 and Vedeld, 2001.

Conclusion

What characterizes the conflicts over Aman is that the staking of the claims takes place in a context in which the institutions of public authority capable of validating those claims are changing due to the process of decentralization. The conflicts over Aman are instructive for studying water governance in Africa because they reveal how water resources are managed within such changing institutional frameworks.

The first situation illustrates how the struggle for rights to the Lake was equally a struggle for authority (see also Lund and Sikor, 2009). The blurred municipal boundaries and absence of statutory decision-making created an opportunity for the contesting parties to gain influence and legitimize the authority of municipalities. The uncertainty about which of the municipalities has the authority to control the use of the lake triggered a conflict between the two who began striving to consolidate their powers, status and income. The municipalities used the argument of having registered citizens in Aman to define their territorial boundaries. In addition, the municipalities referred to events in the past as well as the ability to attract development and investment in public service delivery, whereas the latter became the strongest de facto mechanism for gaining legitimacy and territory in Aman. The different actors' narratives show that boundary-making is a strategic act which draws upon historical practices to create a repertoire of meaning defining who commands control in a changing institutional environment.[18] Furthermore, the process of changing the structures of authority involved in the management of water and natural resources often imply overlaps and discrepancies between customary and statutory rights and institutions (see also Benjamin, 2008). At Lake Aman, an ambiguous co-existence of customary institutions and local government has emerged, which amplifies the existing conflicts of interest between the actors. Although traditional institutions are crucial for the outcome of the decentralization reforms, the challenge of defining role of traditional chiefs in the decentralization process and how to delimit the municipalities remains.

18 See Li, 2000 for similar cases in Indonesia.

In the case study presented in this chapter the mobilization of categories of identity and belonging constitute strategic gambits among the powerful players to gain a winning coalition to control the lake. The questions of who belongs to the community and which community the water belongs to are crucial in an era of decentralization in Mali where struggle over modes of belonging constitute a strategic tool for legitimizing access and excluding "outsiders" from water points and pastures. Like in many other African countries under decentralization, the notion of citizenship has changed from that of uniform national citizenship in the era of independence when postcolonial governments and international donors still saw nation building as an important precondition for development, to an upsurge of struggle over belonging to the "community" without knowing what the community is.[19]

The second situation illustrates how community-based natural resource management approaches promoted under decentralization may enter into conflict with the pastoral livelihood strategy and how pastoralists can easily be excluded under exclusive communal rights (Benjaminsen and Lund, 2003). From this point of view, decentralization is undemocratic, since not all citizens belong to an administratively defined unit and hence have unequal rights to strategic natural resources. In the second situation, the competition for access to resources results in the exclusion from access to water and pastures of mobile herders and "outsiders". So, these groups' access to water is threatened under decentralization. The Fulbe herders are outsiders because they employ land use patterns based upon mobility rather than settlement. Mobility has proven to be well-suited livelihood strategy for the extreme environmental conditions in the Sahel. Nevertheless, the exploitation of Lake Aman demands membership in the community.

The conflicts over Aman present neither an optimistic nor pessimistic view of water governance in Africa. It is a realistic view of resource management in local politics. What happens at the local level is neither a sign of crisis nor collapse. It is the everyday functioning of local politics in an arid environment.

19 See Li, 2000 for similar findings in Indonesia and Ceuppens and Geshiere, 2005 for Cameroon.

Acknowledgements

This chapter is based upon fieldwork conducted for my Ph.D. project which is part of the comparative research program: *Competing for water: understanding conflict and cooperation in local water governance* coordinated by the Danish Institute for International Studies (DIIS). I would particular like to thank my Malian colleagues from Bamako University *Groupe d'Etudes et de Recherche en Droit et Sociologie Appliquée* (GERSDA) Moussa Djiré, Abdoulaye Cissé, Amadou Keita and Anna Traoré. I also owe a great thanks to my research assistants Adama Ongoiba, Mohamed Yayha and Mohamed Arby as well as all the people who patiently participated in interviews.

References

Bakayoko, Issa (2007). La Décentralisation et le Foncier: Les enjeux de la transformation foncière dans la ville de Bamako et son milieu periurbain (Mali). Mainz: Inauguraldissertation, Geschichts- und Kulturswissenschaft der Johannes Gutenberg Universität.

Beeler, Sabrina (2006). *Conflicts between farmers and herders in north-western Mali.* Issue paper 141. London: IIED.

Benjamin, Charles E. (2008). Legal pluralism and Decentralization: Natural Resource Management in Mali. *World Development* 36 (11): 2255–2276.

Benjaminsen, Tor A. (1993). Fuel wood and desertification: Sahel orthodoxies discussed on the basis of field data from the Gourma region in Mali. *Geoforum* 24 (4): 397–409.

Benjaminsen, Tor A. (1997). Natural Resource Management, Paradigm Shifts and the Decentralization Reform in Mali. *Human Ecology* 25 (1): 121–143.

Benjaminsen, Tor A. and Lund, Christian (eds) (2003). *Securing Land Rights in Africa.* London: Frank Cass.

Berge, Gunnvor (2001). Tuareg Notions of Space and Place in Northern Mali. In Benjaminsen, Tor A. and Lund, Christian (eds), *Politics, Property and Production in the West African Sahel: Understanding natural resource management*, p. 182–207. Uppsala: Nordiska Afrikainstitutet.

Bocoum, Aly; Cochrane, Katherine; Diakite, Mamadou and Kane, Oumar (2003). Social inclusion: a pre-requisite for equitable and sustainable natural resource management: Two experiences in Mali. *Securing the commons,* No. 7. London: IIED.

Boone, Catherine (2003). Democratization as political strategy in West Africa. *Comparative Political Studies* 36 (4): 355–380.

Mougin, E., Hiernaux, P., Kergoat, L, Grippa, et al. (2009). The AMMA-CATCH Gourma observatory site in Mali: Relating climatic variations to changes in vegetation, surface hydrology, fluxes and natural resources. *Journal of Hydrology* 375 (1–2): 14–33.

Ceuppens, Bambi and Geschiere, Peter (2005). Autochthony: Local or Global? New Modes in the Struggle over Citizenship and Belonging in Africa and Europe. *Annual Review of Anthropology* 34: 385–407.

Cissé, Augustin; Diabaté, Hubert; Ballo, Desiré; Baltissen, Gerard; Heus, Marleen and Hilhorst, Thea (2003). *Soutenir la Mise en œuvre de la Décentralisation en Milieu Rural au Mali.* Tome 1: thèmes d'actualité. *Amsterdam*: KIT publishers.

Cold-Ravnkilde, Signe. M. (2009). *Struggling for water and pastures in Niger: Natural resource conflict and cooperation in the pastoral areas of the Diffa region.* Saarbrücken: VDM Verlag.

Cotula, Lorenzo, Ed (2006). *Land and Water Rights in the Sahel: Tenure Challenges of Improving Access to Water for Agriculture.* Issue paper 139. London: IIED.

Coulibaly, Aamadi and Hilhorst, Thea (2004). *Implementing Decentralisation in Mali: Two Experiences from rural Municipalities in Southern Mali.* Issue Paper 127. London: IIED.

de Bruijn, Mirjam and van Dijk, Han (1995). *Arid Ways: Cultural Understanding of Insecurity in Fulbe Society, Central Mali.* Amsterdam: Thela Publishers.

de Bruijn, Mirjam and van Dijk, Han (2003). Changing population mobility in West Africa: Fulbe pastoralists in Central and South Mali. *African Affairs* 102: 285–307.

Degnbol, Tove (1999). *State bureaucracies under pressure: a study of the interaction between four extension agencies and cotton-producing farmers in the Sikasso region, Mali.* Ph.D. Thesis, Roskilde University.

Diarra, Tiemoko A. (2008). *Conflict and cooperation in local water governance: Case study of the Douentza district, Mopti region, Republic of Mali preliminary report: hydrological studies.* Prepared for the Danish Institute for International Studies.

Djiré, Moussa (2004). *Mythes et réalités de la gouvernance locale: l'expérience de la municipalité rurale de Sanankoroba, Mali.* Issue paper 130. London: IIED.

Gaasholt, Ole Martin (2009). State decentralization and local politics in Mali, paper presented at the AEGIS Conference, Leipzig June 4–8.

Gallais, Jean (ed.) (1975). *Pasteurs et Paysans du Gourma, la condition Sahélienne.* Paris: CNRS.

Gaudio, Attilio (1988). *Le Mali.* Paris: Karthala.

Hanonou, Eric K. (2009). Slavery and Politics: Stigma, Decentralisation and Political Representation in Niger and Benin. In Rossi, Benedetta (ed.), *Reconfiguring Slavery: West African trajectories*, p. 152–182. Liverpool: Liverpool University Press.

Hetland, Øyvind (2007). *The politics of institutional change in Mali: the democratization, state-society relations and local governance.* Ph.D. dissertation, Department of sociology and human geography university of Oslo.

Kassibo, Brehima (2004). La décentralisation au Mali: état des lieux. *Le Bulletin de l'APAD* 14.

l'Hote, Yann; Mahe, Gil and Some, Bonaventure (2003). The 1990's rainfall in the Sahel: the third driest decade since the beginning of the Century. *Hydrological Sciences Journal* 48 (3).

Leblanc, Marc; Favreau, Guillaume; Massuel, Sylvain; Tweed, Sarah; Loireau, Maud and Cappelaere, Bernard (2008). Land clearance and hydrological change in the Sahel: SW Niger. *Global and Planetary Change* 61 (3–4): 135–150.

Li, Tania M. (2000). Articulating indigenous identity in Indonesia: resource politics and the tribal Sloth. *Comparative Study of Society and History* 142: 149–179.

Lund, Christian and Sikor, Thomas (2009). Access and Property: a question of power and authority. *Development and Change* 40 (1): 1–22.

Marie, Jerome (1975). Le Hombori. In Gallais, Jean (ed.), *Pasteurs et Paysans du Gourma, la condition Sahélienne*, p. 141–161. Paris: CNRS.

Mauxion, Aurélien (2008). Rice Farming Intensification and Political Enterprise in Northern Mali. *Politique Africaine* 110: 153–169.

Moorehead, Richard (1991). *Structural Chaos: Community and State Management of Common Property in Mali.* Ph.D. dissertation, Development Studies, University of Sussex.

NCG (2007). *Etude initial sur la gestion locale de l'eau Mopti-Sikasso, Mali.* Copenhagen: Danida.

Nijenhuis, Karin (2003). Does decentralisation serve everyone? The struggle for power in a Malian village. *The European Journal of Development Research* 15 (2): 67–92.

Oluwu, Dele and Wunsch, James, S. (2004). *Local Governance in Africa: the Challenges of Democratic Decentralisation.* London: Lynne Rienner Publishers.

Ouedraogo, Hubert, M.G. (2003). Decentralisation and local governance: experiences from Francophone West Africa. *Public Administration and Development* 23 (1): 97–103.

PDESC (2005–2009). *Plan quinquennal de développement économique, Social et Culturel de la commune de Hombori.*

Pedersen, Jon (1995). Drought, migration and population growth in the Sahel: The case of the Malian Gourma: 1900–1991. *Population Studies* 49: 111–126.

République du Mali (2002). *Loi n° 02- 006/* du 31 Jan. Portant Code de l'eau.

Ribot, Jesse (2004). *Waiting for democracy- the politics of choice in natural resource management, WRI report.* Washington D.C.: World Resource Institute.

Seely, Jennifer C. (2001). Political analysis of decentralisation: co-opting the Tuareg threat in Mali. *The Journal of Modern African Studies* 39 (3): 499–524.

Thébaud, Brigitte (2000). *Foncier pastoral et gestion de l'espace au Sahel- Peuhl du Niger oriental et du Yagha burkinabé.* Paris: Karthala.

Turner, Matthew D. (2004). Social Conflict and Natural Resources: The Case of Farmer-Herder Relations in West Africa. In Steinberg, M.K. (ed.), *Forests, Fields, and Fish: Politicized Indigenous Landscapes.* Austin: University of Texas Press.

Vedeld, Trond (2001). History, Continuity and Change in Fulani Resource Regimes. In Benjaminsen, Tor, A. and Lund, Christian (eds), *Politics, Property and Production in the West African Sahel: Understanding natural resource management*, p. 117–143. Uppsala: Nordiska Afrikainstitutet.

Zobel, Clemens (2004). Décentralisation, espaces participatifs et l'idée de l'indigénisation de l'Etat africain: les cas de communes maliennes. *Afrique et Développement* 29 (2): 1–25.

Helen Brown*

Social learning and water-resource management in Southern Africa: Can interactive learning trigger changes in water management?

Abstract. The South African and Namibian governments have recently initiated water-sector reform via new legislation (RSA, 1998; GRN, 2004), aiming to increase the equity, efficiency and environmental sustainability of national water resource use. These objectives correspond to those of the integrated water-resource management discourse (Woodhouse, 2008). Institutional reform is proposed as a key mechanism for achieving these objectives. Participatory institutions, such as water-user associations and basin-management committees, are in the process of being formed. These new institutional spaces represent 'communities' of learning (Wenger, 1998; Johnson, 2007), and synergise with the concept of 'social learning' that links collective interaction and learning to concerted action in the collective and environmental interest (Roling & Wagemakers, 1998; Pahl-Wostl et al., 2007; Ison et al., 2007). Drawing on the 'constant comparison' principle of grounded theory (Glaser, 1992), the research described in this article explored social learning within two institutional case studies: the South African Kat River Water User Association (KatRWUA) and the Namibian Kuiseb Basin Management Committee (KuisebBMC). Analysis of qualitative data sources revealed a mismatch between the nature and outcomes of social learning processes within the case studies and the ideals of socially and environmentally sustainable behaviour, as envisaged by both the integrated water-resource management discourse and by the South African and Namibian national Water Acts. In the case studies, social learning, as a process for realising change and achieving goals of

* Department of Geography, Sheffield University. Winter Street, Sheffield, S10 2TN. ggp05hlb@sheffield.ac.uk.

social equity and sustainable human behaviour, was prevented by several hindering factors, including: power relations, personality, precedence, and the past.
Keywords. Social learning, Southern Africa, IWRM, power.

Introduction

Water plays an important role in a number of social, economic and ecological functions (Falkenmark & Rockstrom, 2004). Water in Southern Africa is distributed highly unevenly over time and space (UNEP, 2002), rendering water management problematic. Furthermore, water resources in developing countries are widely held to be in a critical state due to scarcity and depletion (Wescoat & White, 2003). The 2006 World Water Report asserts that this 'water crisis' is "about how we govern access to and control over water resources and their benefits" (UNESCO, 2006: 46). As well as physical water scarcity, the concept of 'social scarcity' is highly relevant to Southern Africa (Ohlsson, 2000), as recent political history permitted the control of and access to water resources by an elite white minority whilst simultaneously denying the black majority population adequate water for economically productive and sanitary livelihoods (Bohensky, 2008). Producing improved approaches to water management is therefore an important issue within the society-environment nexus in Southern Africa.

Social learning and integrated water-resource management

Within the broader environment and development discourse, the concept of sustainable development has transformed water-resource management ideals over the past two decades. 'Integrated water-resource management' (IWRM) has been promoted as a holistic management approach that increases emphasis on the value of ecological systems and social equity, through a shift towards integrated, multi-stakeholder and dialogue-based water-management processes (Savenije & van der Zaag, 2008). Of central theoretical importance to the IWRM discourse is the notion of the 'triple bottom line', which invokes the ideal of decision-making in the social, economic and environmental interest (Hirsch, 2006). The concept of sustainable development is not critiqued in detail here, except to acknowledge the debate surrounding, and increasing criticism of, its relevance and meaning, especially for countries struggling to meet the economic and social needs of existing populations (e.g. Escobar, 1995; Dobson, 1999; Biswas, 2004). Nonetheless, this vision of sustainable development has come to denote 'optimal' natural-resource management.

Whilst understandings and definitions of social learning are multiple and diverse, a recent concept of 'social learning' broadly denotes a managed process of collective learning and interaction, which leads to concerted action (Pahl-Wostl & Hare, 2004; SLIM, 2004; Fazey et al., 2006). This concept of social learning is posited as a mechanism for realising the goals of social and environmental sustainability, which the IWRM discourse promotes (Keen et al., 2005; Ison et al., 2007). Conceptually, social learning therefore links learning with learning-related action, or 'outcomes'.

The concept of social learning, as a management approach, is grounded in an education perspective, which proposes that informed and involved citizens are more likely to enact behaviour that is positive in relation to the group and the natural environment if they understand the need for change, as well as the needs of others. Such understanding is thought to be engendered via participation and interaction centred on a learning ethos (Milbrath, 1989). This notion of learning also draws on Kolb's individual learning cycle, in which action is a reflection of learning (Kolb, 1984; Argyris & Schon, 1996). A social learning approach therefore emphasises that 'learners' be engaged in 'relational tasks' (Craps & Maurel, 2003). Of central relevance to this concept of managed social learning is the assumption that learning within an interactive, task-oriented setting leads to both relational and concerted action outcomes. However, little empirical evidence exists to support this presupposition, especially from developing countries, vis-à-vis the increasing promotion of the approach (Muro & Jeffrey, 2008). This article therefore aims to provide further insight into factors that facilitate or hinder collective learning, as a process leading to concerted action and behaviour change, through an analysis of two empirical case studies of multi-stakeholder social learning processes in South Africa and Namibia: the Kat River Water User Association (KatRWUA) in the Eastern Cape, South Africa, and the Kuiseb Basin Management Committee (KuisebBMC) in the Erongo region of western central Namibia.

The Southern African context

Histories of colonialism and apartheid have shaped power relations and social divisions within contemporary South African and Namibian societies (Buhlungu et al., 2006; Melber, 2007). The historical control of land and water resources by 'elite' social groups, according to racial origin under apartheid, has created socio-economic disparities, as well as differences in the use of, and issues pertaining to, water resources. Since democracy and independence, South Africa and Namibia have respectively undertaken water-sector reform involving policy and institutional change. Revisions to national water policy and legislation encompass the goals of IWRM, as well as objectives to reduce poverty. A key mechanism for the realisation of these objectives is via institutional change, such that new water-management institutions are being created, in order to engender participatory, decentralised water management. Institutions, such as water-user associations and basin-management committees, are expected to be participatory and 'representative' of all the water users within their area of jurisdiction (RSA, 1998; GRN, 2004). In this way, recent policy has created 'spaces' for multi-stakeholder, task-based learning processes (Johnson, 2007). The national institutional frameworks differentiate between water-resource management institutions, and water supply and sanitation service institutions. As water-user associations and basin-management committees are defined as water-resource management institutions, neither the KatRWUA or the KuisebBMC have responsibility for supplying treated water to households. Under the relevant Water Acts, water-user associations are associated with specific management tasks, whilst basin-management committees can potentially perform a wide range of functions. Options include: education, training, data collection, monitoring, implementation of IWRM, serving as an advisory body, infrastructure operation, data analysis, and water use oversight.

The past, the present and the peoples of the KatRUWA and the KuisebBMC

The ratification of the South African National Water Act meant that the former Kat River Irrigation Board (KatRIB), an exclusive institution that represented white commercial citrus farmers, was required to convert to a water-user association and to become more representative in the process. Prior to 1998, water held by the Kat River dam was only made available to scheduled commercial farmers and the Fort Beaufort Municipality. Since 1998, two groups of water-users have increasingly demanded access to this water: the historically marginalised Xhosa population of the catchment and unscheduled commercial farmers, whose previous riparian right to water was overturned by the South African National Water Act. Another 'new' demand on the catchment's water was the mandatory ecological reserve of the National Water Act. Researchers from Rhodes University (RU) used an action research approach to assist with the transformation of the KatRIB by identifying and bringing together representatives of the main water users, in terms of sectoral and volumetric water use. This process essentially redefined the KatRWUA as a representative institution of all water users within the Kat River catchment with the envisaged function of managing water allocations in future (Burt et al., 2007). Although RU researchers were unable to secure the representation on the KatRWUA of all the relevant stakeholders, participants nevertheless included representatives of the main water users, including commercial citrus farmers (scheduled and non-scheduled),[1] emerging commercial

1 Scheduled commercial farmers are those who had historically paid into a scheme run by the Kat River Irrigation Board, which gave them formal rights to water from the

citrus farmers, small-scale farmers, and domestic users, as well as a regional DWAF representative (Burt & Vanderford, 2006).

The KuisebBMC originated in a GTZ-funded project, 'Environmental Learning and Action in the Kuiseb' (ELAK), which was implemented by the Desert Research Foundation of Namibia (DRFN) in 2001. The Department for Water Affairs (DWA) was at that time drafting the new Water Act, which provided for the formation of basin-management committees. Envisaged functions for a basin-management committee are extensive. The KuisebBMC therefore developed as an extension to ELAK, which aimed to foster learning and action towards the sustainable management of the Kuiseb River and associated groundwater resources (see Botes et al., 2003). In this way, the KuisebBMC arose from a learning platform, rather than a pre-existing institution with links to former apartheid-state policies of exclusion and discrimination. KuisebBMC membership is diverse and includes stakeholder representatives of commercial and municipal water users, water-supply institutions, regional authorities, government agencies, research and conservation NGOs, and communal farmers. However, attendance by the latter representatives had notably diminished between the inception of the KuisebBMC and the time that this research took place. The extreme climatic aridity of Namibia drives high physical water scarcity, which is exacerbated by geology type and high evaporation (Mendelsohn et al., 2002). The resultant physical scarcity is reflected in the water-related concerns of stakeholders, which encompass water scarcity, ecological degradation, water supply costs, flooding, and increased regional water demand from the mining industry and urban development (albeit located outside the physical catchment).

Learning and knowledge construction

Information provision and knowledge construction processes are key elements of social learning (Craps & Maurel, 2003; Brown & Pitcher,

Kat River dam and their ability to request releases from it. Unscheduled commercial farmers refer to those who had previously relied on the riparian right of access to water, in order to draw water directly from the Kat River.

2005). Within the KatRWUA, information pertaining to the water resource and the physical and social character of the catchment was presented back to participants in the GIS-based KatAWARE model (Farolfi & Bonte, 2005). Within meetings of the KatRWUA and KuisebBMC, technical and scientific terminology, jargon and concepts were commonly used. Significantly, such information was understood better by historically advantaged participants due to their background of higher formal education, thereby allowing them to use it more effectively to influence the proceedings. In particular, historically advantaged participants of the KatRWUA were able to use such information to justify their current water use and the continued pursuit of livelihoods.

Although facilitating institutions had constructivist goals of allowing and incorporating 'alternative' information and knowledge into the decision-making process, their scope to do so was limited by: the set of highly defined *a-priori* institutional roles; the top-down bureaucracy and guidelines of national state water authorities; and, ultimately, by the strong normative objectives of integrated water-resource management that are imbued in contemporary water-management processes. Information exchange and knowledge construction as collective, 'bottom-up' processes were therefore highly restricted. As a result, the dominant perspectives became equated to collective knowledge within the forums. These findings support Andrew & Robottom's (2005: 73) assertion that, "the insidious nature of the dominant discourses in society is sometimes difficult to recognise and even more difficult to ensure that all types of knowledge and points of view are adequately represented".

Learning can be divided into two conceptual types: the 'simpler' acquisition of external knowledge by the learner, and the more 'complex' internal transition within a learner that involves a change in their perception, beliefs and/or understanding (e.g. Ramsden, 1992). Learning themes that were identified by KatRWUA and KuisebBMC participants included: the different uses and users of water-resources; the need for management on a multi-stakeholder catchment-wide basis; and the importance of water for the natural environment. Collectively, this 'learning' reveals a greater understanding of the complexity of water management on the part of individual, single-sector participants. Research findings also attest to cognitive change on the part of KatRWUA and KuisebBMC participants towards a more integrated and inter-connected socio-ecological system, aligned to the hydrological catchment. Interaction with other stakeholders also produced cognitive change amongst participants, , that other perspectives are valid and that alternative behaviour is therefore rational, if not necessarily 'correct', from personal perspectives. However, the strong parallels of identified learning themes with the capacity-building activities of research and development institutions suggest that power dynamics and interviewer positionality may have skewed responses towards a perceived normative answer.

Furthermore, despite assertions of learning about the need for a balanced approach and consideration of others, such learning was restricted in practice. For instance, in order to justify inclusion in future water allocations, an unscheduled commercial farmer of the KatRWUA asserted that, "farming is sustainable, we have been using the same water all this time and we are still here and there is still water available". This reflection markedly overlooks the inequity of commercial farming practice within the catchment and fails to consider the cumulative impact on water-resource sustainability if all catchment actors had had similarly high water usage requirements.

Interacting

Within both catchments, high inter-stakeholder tension and mistrust had resulted from the historical segregation of ethnic groups and discrimination against non-white citizens. In particular, tensions between upstream and downstream water-users existed. In the Kuiseb River catchment, such tension had been fuelled by perceptions that the commercial farmers of the Upper Kuiseb had been withholding too much water in their dams, thereby preventing it from reaching the Lower Kuiseb (Amakali, 2005). In the Kat River catchment, upstream water-users of the former Ciskei resented their inability to access the water held in the Kat River dam and the lack of water-supply services, as compared with downstream commercial farms. Against this backdrop, research and development agencies played a significant role in bringing disparate social groups together through the respective multi-stakeholder processes, as a result of their perceived neutrality amongst catchment-based stakeholders. In this way, 'bridging' social capital, the links between social groups (e.g. Lehtonen, 2004) were formed via managed social learning processes. Relations between the representatives of different stakeholder groups improved in most cases. For example, many of the stakeholders' attitudes towards other participating stakeholders became more positive, such that most stakeholders lost any pre-existing negative attitudes. Stakeholders that retained or experienced increased negative attitudes were those who had a generally high negative public image, or whose participation was irregular or entirely absent. Face-to-face interaction is therefore considered to be a key driver of relational improvements. Nevertheless, relational changes were also affected by the individual personality and racial identity of stakeholder representatives, as well as individual histories of interaction with formal institutions and associated management processes.

Linking social capital, the vertical links between civil society actors and formal institutions (Lehtonen, 2004), was also generated by the processes. However, the KuisebBMC case was more successful in this sense due to its achievement of greater institutional involvement than the KatRWUA. Yet, although relations had improved, in both cases, tensions had been experienced between the traditional formal water-management authorities and other stakeholders. This finding is attributed to the proposed devolution of decision-making power under IWRM and social learning ideals of democratic decision-making and resource management, both of which differ from previous water-management approaches of South Africa and Namibia. Conversely, adverse relations between the DWAF and commercial farming representatives of the KatRWUA are ascribed to the reclamation of power from the former Kat River Irrigation Board, as the South African National Water Act essentially transferred authority from irrigation boards to the DWAF.

Within both processes, the dominance of historically advantaged and/or institutional members was a significant feature of inter-stakeholder relations. Whilst individual personality played a role, such domination was also made possible by the privileging of scientific and numerical information and the endorsement of 'official' perspectives (e.g. water payment) within the KatRWUA and KuisebBMC forums. Consequently, there was no evidence that the perspectives of the historically advantaged and 'powerful' participants had changed over the duration of the social learning processes. Instead, the perspectives of the historically disadvantaged participants had realigned with them, or else such participants had withdrawn from the process.

Making decisions, taking action and enacting change

Social learning diverges from participation *per se*, by placing social interaction within a context of learning and task-centred activity, in order that learning is then translated into a behaviour, or 'action'. Within the context of IWRM, social learning is envisaged to produce cognitive change that leads to concerted action in the collective social and environmental interest. However, despite identifying improved receptiveness towards the validity of others' perceptions (previous section), unresolved tensions still affected the water allocation process of the KatRWUA. Ambiguous legislation meant that the issue of legality of water use by unscheduled commercial farmers remained unresolved. Scheduled commercial farmers were therefore able to continue rejecting unscheduled farmers' demands for formal entitlements to the Kat River water resource. Allowing equitable allocation to unscheduled farmers would not only decrease the 'allocatable' amount of water but would also allow them to become significant business rivals with the same assurance of supply. Thus, despite the posited cognitive change, that others have valid and rational demands on the water resource, corresponding practical change was restricted if it involved actors potentially losing out economically. In comparison, the ambiguous and, therefore, unthreatening nature of the KuisebBMC forum allowed more cordial relations and a genuine learning ethos to develop.

When questioned about the outcomes of the KatRWUA, one member observed that, "we actually have done nothing but we have carried on [...] we're still around and there is a lot of discussion and debate". Many KuisebBMC participants expressed a similar sentiment:

that the KuisebBMC has not resulted in concerted *related* action by its members or by the wider social groups or institutions that they represent. Overall therefore, whilst both the KatRWUA and the KuisebBMC had made progress in terms of putting relational and administrative structures in place, at the time of this research neither had resulted in concerted collective action towards the broader social and/or environmental change. Despite its lack of concerted action, the KuisebBMC nevertheless provided an effective arena for information exchange and interaction between diverse stakeholders.

Constraints to concerted action

Importantly, both institutions were still developing at the time of this research, thus it is acknowledged that action-based outcomes may still arise in future. However, as recognised by others, several factors hinder the realisation of action by participants of social learning processes (e.g. Leach & Pelkey, 2001; Mostert et al., 2007).

Institutional power relations

The functions of basin-management committees and water-user associations remain strongly subordinate to state agencies, whose approval of a constitution is not only necessary before water-management institutions can obtain legal standing, but who can subsequently revoke such recognition. Relevant caveats in the National Water Acts indicate an unwillingness to devolve real power to non-state institutions and set up a power asymmetry within social learning processes that take place within these institutional forums. The actions and outcomes of the forums were therefore circumscribed largely by formal authorities, rather than by the participants themselves, a feature that is often considered undesirable in terms of the sustainability of voluntary, participatory processes (e.g. UNCCD, 2009).

The decision-making process of the KatRWUA was also strongly influenced by legislative constraints, top-down intervention and the context of the forum. For example, as consultants had classified the Kat River as being of relatively good ecological health, DWAF policy placed constraints on the future health status of the River and, therefore, choices regarding how the resource could be allocated and the water available for socio-economic development. In this way, local decisions about allocation options were constrained by the mandatory human and ecological reserves of national legislation. Furthermore, DWAF retained responsibility for the operation of the Kat River dam, which had a significant bearing on the resource yield as well as the ecological condition of the Kat River. DWAF therefore retained a practical mechanism for vetoing decisions made by the KatRWUA, as well as the power in principle under the terms of the relevant legislation. Ultimately, allocation scenarios and strategies proposed by the KatRWUA constituted 'options' for consideration by DWAF. In this way, the KatRWUA was fundamentally disempowered as a forum where 'real' decision-making could take place. The KuisebBMC suffered from a similar lack of decision-making authority, as observed by a DWA official who asserted that, "at the moment, in this country the Minister for Water decides on everything and the fact that the people who advise him decide". Thus, neither the KatRWUA nor the KuisebBMC practiced devolved decision-making.

Additionally, the decision-making process of the KatRWUA took place within a forum that was facilitated by a research institution which had an extensive history of environment-oriented research in the catchment. Participants were therefore more likely to be motivated to adopt perspectives aligned with environmental goals within researcher-facilitated forums, especially as cooperation with research projects in the Kat River catchment has historically been associated with tangible benefits such as employment and local development (e.g. Perks, 2003).

The legacy of apartheid

Significant differences existed amongst KatRWUA and KuisebBMC members' ability to participate in social learning processes, and to represent their wider stakeholder groups. The apartheid histories of South Africa and Namibia have left wide disparities in educational background, socio-economic status, and self-efficacy. The participation and representation of historically disadvantaged stakeholders were hindered by their financial constraints. Educational disparities further negatively affected their active participation in the KatRWUA and KuisebBMC forums due to the use of scientific concepts and numerical and IT-based information, and predominant use of the English language[2] within them. Finally, the low self-efficacy of many historically disadvantaged participants hampered their ability to represent their stakeholder groups on equal terms within the forums. Despite the oversight and steering of the KatRWUA process by facilitators and a state agency representative, such factors combined to allow a minority of participants to dominate the forum to their advantage, including the prevention of change to the status quo. This 'elite capture' of the KatRWUA process thereby prevented significant change towards participatory objectives of increased social equity.

Desire for economic development

Within the global discourse on water, water use by humans is close to, or in excess of, sustainable limits. Hence new frameworks for decentralised, participatory institutions are perceived as a mechanism for achieving more equitable water allocation and use (Pollard & du Toit, 2008). However,

2 The English language medium is associated with formal education, especially higher education, in both South Africa and Namibia. Consequently, citizens with a higher economic status are more likely to have a higher degree of English language proficiency.

this objective incorporates critical assumptions that the involvement of both historical 'winners' and 'losers' in terms of access to water resources can be secured, and that their interaction will subsequently result in holistic and equitable allocation. Furthermore, in cases where water availability has been fully or over-allocated, equitable water allocation fundamentally entails historically advantaged actors becoming less 'selfish' and relinquishing existing water rights. Given that the majority of livelihoods of the Kat and Kuiseb River catchments rest directly on such rights, it is unsurprising that such actors were unwilling to voluntarily renounce their claims to water resources. Thus, as the allocation activity of the KatRWUA potentially threatened their economic livelihood, representatives of commercial agriculture were strongly motivated to defend their interests within the process. Compared with the economic motivation of KatRWUA members, participation in the KuisebBMC was more strongly driven by awareness that "looking at the bigger picture was necessary". Significantly, a larger proportion of the KuisebBMC membership was of a higher socio-economic class compared with the KatRWUA, suggesting that participation on altruistic grounds is positively influenced by socio-economic status.

Economic considerations not only influenced participation, but also affected water-related decision-making processes. Historically advantaged social actors, both within national contexts and on a wider international level, have set the precedent for affluence and associated living standards to which many historically disadvantaged South Africans and Namibians now aspire. Consequently, public visions for development within the Kat River catchment were consistent with examples of existing socio-economic 'successes', indicating that overcoming tradition and triggering behavioural change is difficult. The desire for economic and social development, and associated socio-economic affluence, largely but not exclusively outweighed local concerns about the natural environment. On a wider national basis, similar populist aspirations to wealth pressure South African and Namibian water authorities to pursue decision making in the interests of social and economic development, which may be to the ultimate detriment of ecological functioning. Achieving 'balanced' decision-making, according to the triple bottom line of the IWRM discourse, is therefore a challenge within the developing country contexts of South African and Namibian due to the wider social, political and economic forces that encourage decisions which engender tangible and immediate socio-economic rewards.

Understanding institutional roles and functions

Institutional activities influenced motivation to participate in social learning processes. As the national institutional framework prevented the KatRWUA from addressing water and sanitation service provision, the KatRVP addressed inequality in the notional sense of a 'right to water' rather than in the practical terms of delivering this entitlement. Hence the process of allocation was of most relevance to those residents for whom basic infrastructural needs were already in place. In this way, the South African National Water Act is inconsistent, as it deters the motivation of the historically disadvantaged to participate in water-resource management processes through constraints on institutional functions, whilst simultaneously advocating their representation. Without such representation, water-management institutions lack legitimacy in the eyes of the state water authorities. Yet state recognition of their legitimacy is fundamental to their future operational viability as water management institutions.

The envisaged roles of the KuisebBMC were significantly more wide-ranging and loosely defined than the KatRWUA. Although intended to incorporate regional variation and to avoid the pitfalls of blueprinting, the lack of a clear function constrained action because, although national legislation *allowed* the KusiebBMC to take on a more active water-management role within the catchment, it was not mandatory. Such an outcome required a positive decision to do so on the part of KusiebBMC members and DWA. This decision had not been taken, partly due to the relational tensions that had arisen from different interpretations of ambiguous policy and legislation regarding the locus of decision-making responsibility and authority at the onset of the process, and partly because many participants were content with the original 'information sharing' focus of the forum. Concerted action by the KatRWUA and the KuisebBMC was further constrained by their lack of available physical, human and financial resources, since, as voluntary membership institutions, they lacked financial self-sufficiency. There is therefore a discrepancy between the social learning discourse, which envisages learning-related action, and the institutional capacity in these specific cases to realise action.

The combined lack of clearly defined functions, limitations to participation and representation, and lack of action to date, meant that both the KatRWUA and the KuisebBMC lacked legitimacy amongst the wider populations of their catchments. Common to both case studies were feelings of mistrust towards the KatRWUA and KuisebBMC. For instance, historically disadvantaged residents feared that involvement in the KatRWUA would lead to demands upon them for water payment. This mistrust was strongly linked to perceptions of the KuisebBMC and KatRWUA as government-aligned 'enforcement' institutions, such that residents within both catchments remained unwilling to engage with them for fear that their current water-related activities would be threatened by their doing so.

Conclusions

Social learning played out at an internalised individual level, but an ultimate concept of action-oriented outcomes was largely restricted by structural constraints and/or subverted by power relations. Power relations are considered central to the explanation of social learning processes and their outcomes, as they limited local capacities of agency, both on the part of historically disadvantaged individuals within the processes and on the part of the locally based institutions to pursue concerted action. These findings synergise with conceptualisations that power relations are: inherent amongst social actors; multi-scale, and; latent or explicit (Lukes, 2005). Power imbalances between social actors within the case-study processes resulted, variably, from differentials in resource access, personality, institutional affiliation and socio-economic status. Ultimately, all these factors were influenced significantly by recent national histories. This interpretation supports Waalewijn et al. (2005), who conclude that decentralised participatory water-management in the South African context cannot be representative or free from power relations until wider society is correspondingly more equitable. Such power differentials highlight the importance of skilled and independent facilitation of social learning processes if elite capture of the process is to be avoided. The need for facilitation of social learning processes is especially relevant to their use in developing countries, where populations retain high intra-social disparities in socio-economic status.

Whilst power relations were influential in governing institutional action and individual behaviour, the political, economic, physical and social characteristics of the respective catchments and countries also affected the play out of social learning processes. Despite identifying learning and cognitive changes amongst KatRWUA and KuisebBMC participants, the subsequent failure to realise intended behavioural outcomes, either individually or collectively, indicate that other factors govern behaviour. Both wider structural constraints (e.g. hierarchal institutional frameworks, ambiguous legislation and wider economic structures) and internal constraints (e.g. motivation and self-efficacy) played a significant role in preventing subsequent action. Multi-scale economic factors were exceptionally influential, as they negatively influenced the motivation and desire of stakeholders to participate in social learning processes and, furthermore, to pursue concerted action in the collective and environmental interest. The practical implementation of social learning approaches is therefore problematic.

This analysis of social learning processes generates cause for optimism due to their ability to foster goodwill and a spirit of cooperation amongst diverse social actors. Such relational outcomes are highly relevant to the Southern African context, where social capital is relatively low for historical reasons. However, the ability of powerful stakeholders to dominate and influence proceedings to their own material advantage also provides cause for pessimism, as such outcomes are likely to negatively affect the sustainability of such processes. Thus, whether social learning processes can deliver real action and change will determine their future relevancy.

Acknowledgements

I am highly indebted to the members of the Kat River Water User Association and the Kuiseb Basin Management Committee for letting me attend their meetings, giving up their time for interviews and for patiently answering my questions. Rhodes University Geography Department and the Desert Research Foundation of Namibia provided valued support and assistance whilst I was conducting the fieldwork upon which this article is based. Finally, financial assistance from the UK Economic and Social Research Council, the University of Sheffield and the Dudley Stamp Memorial Fund enabled this research to be carried out.

References

Amakali, Maria (2005). Intra-state conflict resolution between local communities and central governments – Namibian cases. Conference paper *Value of water: Different approaches in transboundary water management*, 10/03/05, Koblenz, Germany.

Andrew, J. and Robottom, I. (2005). Communities' self-determination: Whose interests count? In Keen, Meg; Brown, Valerie and Dyball, Rob (eds), *Social learning in environmental management: Towards a sustainable future,* p. 63–77. London: Earthscan.

Argyris, Chris and Schon, Donald (1996). *Organisational learning 11: Theory, method and practice.* Reading, MA: Addison-Wesley.

Biswas, Asit (2004). Integrated water resources management: A reassessment. A water forum contribution. *Water International* 29 (2): 248–256.

Bohensky, Erin (2008). Discovering resilient pathways for South African water management: Two frameworks for a vision. *Ecology and Society* 13 (1).

Botes, Andre; Henderson, Judith; Nakale, Tufikifa; Nantanga, Komeine; Schachtschneider, Klaudia and Seely, Mary (2003). Ephemeral rivers and their development: Testing an approach to basin management committees on the Kuiseb River, Namibia. *Physics and Chemistry of the earth* 28: 853–858.

Brown, Valerie and Pitcher, Jennifer (2005). Linking community and government: Islands and beaches. In Keen, Meg; Brown, Valerie and Dyball, Rob (eds), *Social learning in environmental management: Towards a sustainable future,* p. 123–145. London: Earthscan.

Buhlungu, Sakhela; Daniel, John and Southall, Roger and Lutchman, Jessica (eds) (2006). *State of the Nation: South Africa 2007.* Cape Town, South Africa: HSRC Press.

Burt, Jane and Vanderford, Paul (2006). *Identifying stakeholders for the Kat River catchment WUA*. Pretoria: Water Research Commission.

Burt, Jane; McMaster, Alistair; Rowntree, Kate and Berold, Robert (2007). *Local institutions for water governance: The development of a Water User Association and Catchment Forum in the Kat River Valley, Eastern Cape.* Pretoria: Water Research Commission.

Craps, Marc and Maurel, Pierre (2003). *Social Learning Pool of Questions. An instrument to diagnose Social Learning and IC tools in European river basin management.* WP2/WP3 report of the HarmoniCOP project. Brussels: HarmoniCOP.

Dobson, Andrew (1999). *Fairness and Futurity: Essays on Environmental Sustainability and Social Justice.* Oxford: Oxford University Press.

Escobar, Arturo (1995). *Encountering development: The making and unmaking of the third world.* Princetown, N.J.: Princetown University Press.

Falkenmark, Malin and Rockstrom, Johan (2004). *Balancing water for humans and nature: The new approach to ecohydrology.* London: Earthscan.

Farolfi, Stefano and Bonté, Bruno (2005). *KatAWARE: A negotiation-support tool for participatory water resource management in the Kat River Valley (Prototype model).* WRC project no. K5/1496. Grahamstown: Institute for Water Research, Rhodes University.

Fazey, Ioan; Proust, Katrina; Newell, Barry; Johnson, Bill and Fazey, John, (2006). Eliciting the implicit knowledge and perceptions of on-ground conservation managers of the Macquarie Marshes. *Ecology and Society* 11 (1): Article 25.

Glaser, Barney (1992). *Emergence vs. forcing: Basics of grounded theory analysis.* Mill Valley, CA: Sociology Press.

GRN (2004). *Namibian Water Resources Management Act* (Act No. 24 of 2004). Windhoek: Government of the Republic of Namibia.

Hirsch, Philip (2006). Water governance reform and catchment management in the Mekong region. *The Journal of Environment and Development* 15 (2): 184–201.

Ison, Ray; Roling, Niels and Watson, Drennan (2007). Challenges to science and society in the sustainable management and use of water: investigating the role of social learning. *Environmental science and policy* 10: 499–511.

Johnson, Hazel (2007). Communities of practice and international development. *Progress in Development Studies* 7: 272–290.

Keen, Meg; Brown, Valerie and Dyball, Rob (eds) (2005). *Social learning in environmental management: Towards a sustainable future.* London: Earthscan.

Kolb, David (1984). *Experiential learning: Experience as the source of learning and development.* New Jersey: Prentice Hall Inc.

Leach, William and Pelkey, Neil (2001). Making watershed partnerships work: A review of the empirical literature. *Journal of Water Resources Planning & Management* 127: 7.

Lehtonen, Markku (2004). The environmental-social interface of sustainable development: capabilities, social capital, institutions. *Ecological Economics* 49: 199–214.

Lukes, Steve (2005). *Power: A radical view.* Basingstoke: Palgrave Macmillan.

Melber, Henning (2007). *Transitions in Namibia: Which changes for whom?* Uppsala: The Nordic Africa Institute.

Mendelsohn, John; Jarvis, Alica; Roberts, Carole and Robertson, Tony (2002). *Atlas of Namibia: A portrait of the land and its people.* Cape Town: DP Publishing.

Milbrath, Lester (1989). *Envisioning a sustainable society: learning our way out.* Albany, NY: University of New York Press.

Mostert, Eric; Pahl-Wostl, Claudia; Rees, Yvonne; Searle, Brad; Tabara, David and Tippett, Joanne (2007). Social learning in European river-basin management: barriers and fostering mechanisms from 10 river basins. *Ecology and society* 12 (1): Article 12.

Motteux, Nicole (2003). *Evaluating people-environment relationships: developing appropriate research methodologies for sustainable management and rehabilitation of riverine areas by communities in the Kat River Valley, Eastern Cape Province, South Africa.* PhD thesis, Rhodes University, Grahamstown, South Africa.

Muro, Melanie and Jeffrey, Paul (2008). A critical review of the theory and application of social learning in participatory natural resource management processes. *Journal of Environmental Planning and Management* 51(3): 325–344.

Ohlsson, Leif (2000). Water Conflicts and Social Resource Scarcity. *Physics and Chemistry of the Earth (B)* 25 (3): 213–220.

Pahl-Wostl, Claudia and Hare, Matt (2004). Processes of social learning in integrated resources management. *Journal of Community and Applied Psychology* 14 (3): 193–206.

Pahl-Wostl, Claudia; Sendzimir, Jan; Jeffrey, Paul; Aerts, Jeroan; Berkamp, Ger and Cross, Katherine (2007). Managing change towards adaptive water management through social learning. *Ecology and society* 12 (2): Article 30.

Perks, Camilla (2003). Kat River Valley rescued. *Rhodos Staff Newsletter* 15(18): 12th December 2003.

Pollard, Sharon and du Toit, Derek (2008). Integrated water resource management in complex systems: How the catchment management strategies seek to achieve sustainability and equity in water resources in South Africa. *Water SA* 34 (6): 9.

Ramsden, Paul (1992). *Learning to Teach in Higher Education.* London: Routledge.

Roling, Niels and Wagemakers, Maria (1998). *Facilitating sustainable agriculture; participatory learning and adaptive management in times of environmental uncertainty.* Cambridge: Cambridge University Press.

RSA (1997). *Water Services Act, Act 108 of 1997.* Pretoria: Government of the Republic of South Africa.

RSA (1998). *The National Water Act, Act 36 of 1998.* Pretoria: Government of the Republic of South Africa.

Savenije, Huub and van der Zaag, Pieter (2008). Integrated water resources management: Concepts and issues. *Physics and Chemistry of the Earth* 33: 209–297.

SLIM (2004). *The role of learning processes in integrated catchment management and the use of sustainable water.* SLIM Policy Briefing 6. Milton Keynes: Open University.

UNCCD (2009). *Participatory development: A bottom-up approach to combating desertification.* Fact sheet no. 6. Bonn, Germany: UNCCD.

UNEP (2002). *Africa Environment Outlook. Past, present and future perspectives.* Hertfordshire: Earthprint.

UNESCO (2006). *Water: A shared responsibility, World Water Development Report.* Paris: UNESCO.

Waalewijn, Pleter; Wester, Philippus and van Straaten, Kees (2005). Transforming river basin management in South Africa: Lessons from the lower Komati River. *Water International* 30 (2): 12.

Wenger, Etienne (1998). *Communities of Practice: Learning, Meaning, and Identity.* Cambridge: Cambridge University Press.

Wescoat, James and White, Gilbert (2003). *Water for life: Water management and environmental policy.* Cambridge: Cambridge University Press.

Woodhouse, Philip (2008). Water rights in South Africa: Insights from legislative reform, in *Brooks World Poverty Institute Working Paper,* ssrn.com/abstract=1265588 (accessed 21 Jan 09).

Daniel Alvarenga*

How does the river flow? The Mphanda Nkuwa project and China's impact on Mozambique and Southern Africa's water and development

Abstract. This paper looks at an important dimension of China's engagement in Mozambique – its infrastructure sector and the Mphanda Nkuwa dam set to be built on the Zambezi River. Its envisaged impact on the water profile of Mozambique is noteworthy.
Keywords. Mphanda Nkuwa, China, Mozambique, Development.

Introduction

Looking at the backdrop of negotiations towards building Mphand Nkuwa is a telling window into the intricacies of China's engagement in the country. By looking more closely at the political nuances of the negotiation process, we hope to shed some light on the dynamics behind the project and pave the way for a more informed discussion regarding the environmental impact of the infrastructure project.

Literature on the local micro-political economy will be relevant to assessing the expected impact of the dam whilst also contextualizing the exchange of high-level diplomacy concerning the construction of the dam as it unfolds.

The background concerning the impacts of mega-projects, economic development and the implications of the strategic options of the Mozambican government will be taken into account.

* Former analyst Centre for Chinese Studies at Stellenbosch University (South Africa); daniel.alvarenga@yahoo.co.uk; <www.cs.org.za>.

After providing a history of the dam, data from the negotiation process will be analysed for important tendencies regarding Mozambique's economic policy-making and that of Southern Africa. The paper concludes with a broad debate on how the dam may positively and negatively affect Mozambique's development.

The story of the dam – genealogy of Mphanda Nkuwa

Mphanda Nkuwa is on the verge of becoming the largest infrastructure scheme in Mozambique involving Chinese investment. The construction of the 1,500 MW capacity dam on the Zambeze river was expected to start in 2009 and to be completed in 2014 but has suffered delays and is now expected to begin in 2011. Its site will be in the lower Zambezi Area, 70 km south of the famous Cahora Bassa dam. A small description of the Zambezi River is also called for. It is one of the longest rivers in Africa and has already been the target of many similar projects with 30 dams in its basin that extends from Angola to Mozambique.

Two of the largest dams are those of Cahora Bassa, in Mozambique, and the Kariba dam in Zimbabwe. The river valley is estimated to hold about 72 million people, of which about 80% are dependent on the river for agriculture and fishing. The total cost of the structure was initially estimated to be US$2.3 billion, of which US$1.1 billion are for the dam itself and US$1.2 billion for the transmission lines but this estimate has since increased to US$2.9 billion. The initial promoters were the Ministry for Mineral Resources and Energy of Mozambique and NEPAD. At an early stage in the negotiation, the European Investment Bank and the World Bank had expressed interest in financing the project but were dissuaded by the risk-analysis contained in the 2001 feasibility study carried out by UTIP "Unidade Técnica de Implementação dos Projectos Hidroeléctricos" (Technical Unit of the Implementation of Hydroelectric Projects) (Meguigy, 2002).

Chinese involvement undertaken through the effort of the China Exim-Bank that has encouraged the dam's construction from the start and is set to become one of the key financiers of the project. An MOU was signed between the Mozambican government and Exim Bank in April 2007. The deal however has now been approved by the government and the construction contract has been given to Brazil's Camargo Corrêa with Electricidade de Moçambique and Energia Capital as partners. The first agreement between Mozambique's government and Camargo Correa was signed in September 2007, after the MOU between the Mozambican government and China Exim Bank.

Ownership of the dam will be divided 20 percent by the state-run Electricidade de Moçambique (EDM) and 80 percent by a 50–50 joint venture between the local company Energia Capital and Brazil's Camargo Correia. According to the government, these three companies will provide 30 per cent of the project investment whilst the rest will come from financial institutions, with Exim Bank seen as a likely partner according to local reports (Portuguese News Network, 2011; Reuters, 2010; All Africa, 2010).

Camargo Correa sees Mozambique and Mphanda Nkuwa as its litmus test (Lusa, 2008) for proving itself as a top multinational company. The project's outcome is seen as important for the fate of the company's strategy in the medium-run, as Camargo is hoping that the venture will project the company onto the global stage. China Exim Bank has pushed for the construction of the dam from an early stage and is a very natural choice as a financial partner. The entire process would not have arrived at this stage without its action and lobbying. At a previous stage, the World Bank was expected to request partial funding (Soren Ambrose, 2008) and a partnership between Exim-Bank and the World Bank was originally contemplated. Such might not seem very significant at a first glance but, actually, the very fact that those two institutions were open to such a partnership revealed a significant shift in their Africa strategy. Equally novel is the form of mutual perceptions and interactions between two of the major financial institutions operating in Africa.

Cooperation and coordination between the Exim-Bank and the World Bank would go a long way toward improving the efficient application of financial resources on the continent as well as exponentially increasing the know-how and capacity for large-scale financing schemes such as Mphanda Nkuwa. It would also set up a system of checks and balances between the two institutions, even if limited to occasional projects, and keep a healthy, desirable and practical dialogue between China and the Western Actors represented by the World Bank regarding their relationships with Africa.

From the Mozambican government's perspective, Luísa Diogo expressed in 2007 (*Jornal Notícias de Moçambique*, 2007 A) that negotiations should run their natural course and that the government would not rush the project. The Prime Minister expressed some worry that a project of such dimension might consume the majority of international investments in Mozambique and leave other projects and areas forgotten.

Armando Guebuza in turn, the country's head of state, saw Mphanda Nkuwa as an important step towards fulfilling Mozambique's energy production potential which he claimed to be of more than 12,000MW. Within the framework of this wider ambition, the government has set up a technical institute responsible for training in the areas of supply, transport, distribution and administration in the energy sector (*Jornal Notícias de Moçambique*, 2007 B).

This is an instance of how China's interest in the infrastructure sector also managed to encourage Mozambique's governing institutions to improve, reform and expand to make the most out of investment opportunities.

Finally, it is also worth noting that there are three possible forms of ownership in the electricity industry:

1) direct government ownership;
2) government-owned corporations;
3) privately owned corporations (Yi-Chong, 2004: 31).

Along the same lines, there are four possible power structures in the industry:

1) vertical/horizontal monopoly;
2) purchasing agency model (allows for competition in generation);
3) wholesale competition (competition in generation and distribution);
4) retail competition (competition in generation, distribution and consumers) (Yi-Chong, 2004: 2).

The Mphanda Nkuwa Dam consortium incorporates a mix between a government owned corporation and a privately owned corporation. When it comes to the particular structure of Mozambique's power industry, government has allowed wholesale competition to take place, with different companies tendering for the construction of energy production facilities as well as distribution lines connecting to the national grid. There is, however, no competition in retail, production and distribution since, *Electricidade de Moçambique* holds the monopoly on the consumer market.

Chinese pragmatism

A contextualized analysis of the dam's negotiation shows notable signs of Chinese pragmatism. Elsewhere in Africa, the absence of power infrastructures has led the Chinese to develop their own sources of power for key projects such as raw material extraction. So, although China's motives in financing Mphanda Nkuwa were similar to those of other African projects, the way it has gone about implementing them is distinct.

A comparison with the Merowe Dam in Sudan will reveal the extent to which the dynamics of China's interest and practices in Mphanda Nkuwa are an exception. Merowe is a classic example of Chinese practices in the continent. It is a US$1.8 billion project whose negotiation, conditions and construction are similar to Mphanda Nkuwa. Merowe's construction has been carried out by a consortium of two European and two Chinese companies and financed by Middle Eastern entities in partnership with Exim Bank.

Its dimensions are similar to those of Mphanda Nkuwa, being projected to produce 1250 MW of power (Burke, 2007). As with the Nkuwa Dam, the political debates regarding Merowe have also revolved around the displacement of people as well as the reliability (and at times even existence) of feasibility studies. The Sudanese government is expected to have relocated more than 50,000 people. In the negotiation process of Merowe dam, China managed to win a second contract for power towers to transport electricity to Khartoum and Port Sudan, with Harpin-Jilin and CCMD comprising a joint venture respectively for US$10 million. According to Askouri (2007: 76), an "inside source said Sinohydro (engineering) told Zesco (Zambia Electricity SupplyLimited) that they wanted to see a site assessment focused only on economic factors". This has been one of the similar critiques that Mphanda Nkuwa has been subject to, i.e. that it is overly concerned with economic viability for the investors and disregards issues of sustainability and the environment.

It is apparent that, when carrying out a project like Mphanda Nkuwa, China often contracts its own domestic companies such as Sinohydro.

However, Mphanda Nkuwa's case is, to some extent, unique. Firstly, although the opening of the Mozal II aluminium smelter has increased overall industrial demand for electricity, Mozambique already generates an energy surplus, even if the consumption of families greatly expands in the near future. These circumstances are likely to continue until the majority of the population actually gains access to electricity and consumption picks up.

Secondly the economic diplomacy of Mozambique's government and the valuable know-how of the Brazilian engineering company have pushed China, or more precisely, one of its key foreign policy-making actors, Exim-Bank, to adjust its traditional *modus operandi*. This time, the model is more similar to the Merowe Dam scheme: China Exim-Bank seems likely to have third party involvement in a kind of deal that is usually a bilateral agreement between two governments and relies upon Chinese financing of a Chinese-constructed project.

This project may just possibly prove to be an exception proving the rule that China's preference for vertical integration of its investments will not change. It does, nonetheless, send out an important signal that China's investment and loan packages are likely to become more flexible and innovative. There are two reasons as to why this development is not surprising. Firstly, China has an unremitting tendency to be pragmatic in its foreign affairs and economic diplomacy, Mphanda Nkuwa being a good example. Secondly, as its engagement with the international market increases, the business practices of China Exim-Bank will, just like the practices of other Chinese MNCs, tend to transcend the most immediate interests of the PRC State Council and acquire a fluid and more autonomous dynamic of its own.

Hence, the Mphanda Nkuwa case stands as a reminder that both the grand and the minor narratives of China's engagement of Africa are anything but static and can no longer be encapsulated in the maxim of "Chinese financing for Chinese-built and engineered projects". The process leading up to the construction of Mphanda Nkuwa has shown signs of pragmatism and flexibility on behalf of the actors defining and executing China's engagement with Africa on the Chinese side.

The regional dimension of China's strategic approach

A major electricity crisis for South Africa, the major power consumer in the region, increases the significance of the Mphanda Nkuwa dam. Eskom, the South African national energy company, is expanding its investment strategy and is now negotiating to buy an additional 250MW of electricity per day from Mozambique's *Hydroeléctrica de Cahora Bassa* (HCB). South Africa has even considered splitting into two time zones for energy-saving purposes (Yeld, 2008). The stage is set for the country to gradually become a regional dynamo and also for China to play a role. The output of the dam is closely tied to regional energy security issues. South Africa is set to be the main purchaser of the electricity produced by the dam together with Malawi and Zimbabwe. Regional energy interdependence exists to the extent that Mozambique is expected to become the leading regional supplier in the region (Engineering News, 2008).

Mozambique is, in fact, connected to the Southern Africa Power Pool (SAPP). However, South Africa and Zimbabwe as well as Mozambique still face serious distribution difficulties, particularly in rural areas. This is reflected both in an ADB report on Mozambique in which "64% of the sampled companies have identified energy as being a major problem with 17 power cuts registered per month". This points up two interlinked paradoxes: one between Mozambique's energy production potential and the actual energy production/distribution; and another between energy revenues and the internal capacity of Mozambique for generalizing its benefits across the population and businesses.

Considering other mega-projects in the region, a dam such as Mphanda Nkuwa seems to be a relatively safe economic investment given the current regional context regarding energy production and consumption. This is so because "electricity is not a normal good that consumers can pick and choose at a market where there is a free exchange system"; pricing is inelastic (despite a possible economic downturn, demand is still likely to remain at a high level as long as a broad distribution network is in place).

The vast majority of commercial energy consumption in the region is generated by electricity rather than petroleum or coal. There are also low levels of internal energy consumption and access rates in Mozambique vis-à-vis other countries in the region.

The Mphanda Nkuwa project is increasingly regional, even in regard to its country-specific investments. After noting how China takes into account regional demand and dynamics when evaluating economic viability, there is a clear regional dimension to China's strategy, even in such a one-off project. This is due to the fact that producing energy for internal consumption is not yet commercially viable but also that investors can take advantage of Mozambique's membership in the SAPP. The regional dimension of China's involvement in the energy sector in general and in projects such as Mphanda Nkuwa in particular are examples of how China's involvement manages to foster absolute gains for those involved. In this particular instance, by linking its investments to Mozambique's regional capacity and its connection to SAPP, China will contribute to the economic integration of regional blocs such as SADC, of which Mozambique is also a part. China is promoting political and economic integration between Mozambique and its neighbors by means of spillover effects borne out of the need to find effective solutions for new challenges and opportunities.

How will the dam affect Mozambique's development?

Mozambican politics have long been characterized by the hegemony of decision-makers in the capital over those in local government. However, local government has historically worked around a *de facto* incapacity of the national authorities to effectively carry central governmental decisions all the way to isolated peripheral regions (Cuereneia, 2001).

Thus a closer analysis is required of the dynamics between central and local government in the Zambeze region, where the dam is to be constructed. The area surrounding the building site is predominantly rural, with isolated families and villages spread out in the mountainous areas laboring in local subsistence agriculture, cattle herding and traditional crafts such as carpentry and pottery.

It would be an oversimplification, however, to imagine the local political network as being organized around a simple subsistence economy. In fact, there are considerable inner-group and inter-group power inequalities which add another dimension, locking some of the subsistence farmers into forms of dependency (Morrissey, 2006).

The presence of a project of such dimension and nature contrasts sharply with the immediate picture of the local micro-economic dynamics. This contrast is an ironically fitting allegory to the challenges Mozambique's macro-economic structure faces. Mozambique's government has previously been accused of claiming credit for almost "miraculous" growth rates which are sometimes overwhelmingly dominated by the contribution of megaprojects.

Mphanda Nkuwa constitutes one of these very projects. This critique arises from studies that have broken down the picture of how different strata and socio-economic groups have derived highly differing benefits from these high growth rates. Neither is this critique straightforward, however, as the true nature and impact on the poverty of Mozambique's economic development remains widely debated (Arndt, 2007).

The main potential risks in the construction of the dam encompass, first of all, its location in a potentially seismically risky area. There are not only the traditionally risks related to "reservoir induced seismicity" but also chronic lapses in the collection of seismic records. Mozambique is "in vicinity of Nubia-Somalia plate boundary and straddles a highly active fault zone called the shire though" (Lemos & Ribeiro, 2007: 67). In addition, besides floods, draught remains a primary concern for the local population. The arrival of the dam will change the patterns of river flow with uncertain risks and consequences for those who have traditionally relied on it.

A useful way of looking at what are, for the purposes of this project, the most relevant socio-economic aspects of the building of the dam is by dividing them into two categories as *Justiça Ambiental* and IRN (2006) did: those related to construction and those caused by dam operation. In regards to the former, the very action of building the dam entails specific changes for the populations residing in the local construction site.

Characterized by a traditional absence of the central state in its socio-political organization, these traditional groups will be subject to a significant wave of immigration from urban areas to work in the mega-project. Critics claim that, despite greater economic and commercial activities and new job opportunities, crime and HIV/AIDS rates are also likely to increase. To that, one must add the obvious onus of resettlement with the first estimates pointing to a number of up to 200,000 people to be relocated.

This paper distinguishes between the risks and challenges related to dam operation and those concerning the dam construction and applies them to the realms of economic diplomacy especially in regards to the implications for local populations.

Reports such as those of *Justiça Ambiental* and IRN (2006) defend that the dam will lead to greater severity in droughts with subsequent humanitarian catastrophes and food security problems due to lower and less stable levels of water in the Delta. A similar train of thought argues that China Exim Bank's involvement in Nkuwa has removed the pressure on the Mozambique government to improve social and environmental assessments (Lemos & Ribeiro, 2007: 68).

Justiça Ambiental and IRN offer as alternatives a set of other options such as micro-hydro, solar panels, windmills, biofuels from sugarcane waste and even improved cooking stoves (IRN, 2006). These can undoubtedly be important complements to hydroelectric power but some doubts can be posed regarding the extent to which these constitute viable alternatives if the objective is to commercialize and export electricity in large quantities.

In contrast to this antagonistic position towards the dam by a considerable section of organized civil society, the government and other economic agents defend the exact opposite (*Jornal Notícias de Moçambique*,

2007 A). The project is seen by authorities not just as being economically sound but it also as offering the advantage of greater control over the river flow and flood prevention. In fact its economic viability is not certain, and it seems that the dam may be particularly vulnerable to climate change and natural catastrophes.

A case in hand is the Kariba dam upstream which has been operating at 14 percent of capacity due to recurring droughts. Taking into account climate change also assumes particular relevance as Southern Africa is one of the regions in the world predicted to feel the greatest impact in the form of desertification and droughts in the medium/long term (Brown et al., 2007; Eriksen et al., 2007).

These debates and challenges point once again to the importance of going beyond the mere quantification of the potential impacts and implications of the dam. Any macro-economic and macro-political analysis needs to be complemented by an understanding of the micro-economics and the micro-politics as well as outside-in thinking.

In itself, the project holds some very interesting overlapping challenges when it comes to transnational issues such as those of social and environmental responsibility issues and even taxation (Bucuane & Mulder, 2007). The most important is for the income from electricity exports to be successfully distributed and re-invested productively into the economy.

The sheer dimension of the dam means that once the project goes past its tendering phase and kicks off, Mphanda Nkuwa will become a flagship in Mozambique's energy strategy. It is therefore of the utmost importance that the Mozambican government, because of its privileged position as income collector, commit itself to ensuring that it benefits as many people as possible.

Just as Soderbaum and Taylor's identified in their analysis of the Maputo Development Corridor (MDC) initiative and the role of the state, there is the risk that flaws, disregard and state-elitism in a mega project like Mphanda Nkuwa may "reinforce the role of the state as a transmission belt for international capital, rather than as a facilitator for development" (Soderbaum & Taylor, 2001: 676).

In brief, the double conceptualizations of the state as a transmission belt and as a facilitator for development bring the state and, more precisely, the government into a useful analytical dichotomy. It presents a variation from the state as an actor that operates as a transmission belt, this concept presenting an understanding according to which the state carries out the task of being the "disciplining spokesman of transnational capital".

This contrasts with Cox's (1996) view that the state should act as a "buffer" protecting the domestic economy from harmful exogenous influences. To Cox's ideal type one can add not just the protective function of the state but also its role in potentiating positive exogenous influences. These two contrasting conceptualizations point out to how the government of Mozambique must, at the end of the day, be more aware and active regarding the concrete impacts China is having on its economic development, constituting both threats and opportunities.

The MDC's paradigm was characterized by "short-term, capital-intensive and large scale investment, with a minimal degree of state involvement." Some flaws in this paradigm seem to have been repeated with Mphanda Nkuwa.

As in the case of the MDC, and other mega-projects, local populations have not truly been consulted before or during the process of planning and execution of the projects nor were they "integrated into the decision-making" (Soderbaum & Taylor, 2001: 688).

The dam remains nonetheless a major macroeconomic "chess-move" the impact and benefits of which have yet to be broken down and analysed in order to identify those which have contributed the most to economic development performance – local poor rural classes. For this to happen, however, the government will have to properly manage the taxes collected from the newly exported electricity from the dam. Mozambique has adopted a strategy concerned with "increasing the profile and credit rating of the region" (Soderbaum & Taylor, 2001: 686; Business Map, 2000: 37).

The country needs to avoid the over-reliance on deals "between the political elite and transnational capital" that merely benefit these two actors. There must be a distinction between a government that transfers its responsibilities by giving full autonomy to a company to carry out all the "grunt" work, i.e. managing risks, and a government that provides incentives and stimulates new and innovative investment models. The Government of Mozambique needs to distinguish between utter carelessness and *l'état stratége.*

As long ago as 2001, Soderbaum and Taylor were already pointing out to how there were no institutions able to "facilitate dialogue with bottom-up forces in society" (Soderbaum & Taylor, 2001: 692). Despite the positive openness of China to adapt and change its investment models as required, the dialogue remains restricted to the elite forces in society. In fact, the very objective of producing 12,000 MW mentioned by Armando Guebuza and referred to in the previous section of this paper has still fallen short of being achieved and much of the energy produced by mega-projects stands to benefit, above all, other mega projects.

Examples of such projects include the heavy sands exploration of Chibuto and Moma as well as Mozal III aluminium smelter, businesses that depend on reliable and cheap energy supplies. This creates a case of a double economy and a closed elitist circle of investment supporting the arguments pointing to the paradox of growth vs. inequality.

Most strikingly, Mozambique's strategic option of trying to produce energy in greater quantities for export highlights an extreme paradox when one looks at its internal consumption: what will a dam producing 2500 MW for export do to a country in which households still rely on "biomass in the form of charcoal and firewood for 80–90% of its internal energy consumption" (Ghazvinian, 2008)?

Conclusion

This paper has reviewed the tendering process of Mphanda Nkuwa. It has found that the genealogy of the dam shows significant signs of pragmatism from Chinese actors in Africa and particularly the Exim-Bank which has demonstrated increasing openness to reinventing its partnership and investment models. That said, these changes also come at a time when more autonomous actors plan out the investments and projects taking into account a regional framework. Finally, in a nutshell the answer to the question "who will benefit from the dam?" remains to be answered by the government which will collect the revenues from electricity exports, since the economic benefits from the construction of the dam itself are short and circumstantial.

There is still a need for the Mozambican government to evolve from being a "transmission belt" for China's transnational capital to a "facilitator of development". Mozambique governance structures should not simply allow China to displace Western partners and obtain short-term profit-seeking deals, but actually tap and adapt China's commitments on an individually tailored, project-by-project basis.

The Mphanda Nkuwa project has not been without its critics. Serious questions have been raised regarding the potential displacement of people. Furthermore, there have been criticisms regarding the reliability of the feasibility studies that have been conducted. IRN and the NGO *Justiça Ambiental* are two of the most vocal groups demanding further studies. Some of their representatives even lobbied Chinese stakeholders during the ADB meeting in Shanghai in 2007 (International Rivers Network, 2007).

Mozambique must now also look at starting to make the most out of investments that are already there, by creating incentives for these businesses, previous investments and mega-projects to stay and expand. This would be a new and welcome trend given recurrent suggestions for a greater focus on capacity-building instead of simple carefree liberalization. If construction comes about, it will be a political decision if Mphanda Nkuwa is economically successful and if it will benefit a wide constituency or not. The nature of economic impact is similar to that of other mega-projects and the same development benefits and limitations apply. Meanwhile the government of Mozambique will retain the last word in the developmental impact of the dam which will depend on the re-distribution and re-investment of the significant revenues that will be collected from exporting the newly produced energy.

References

ADB & ADF (2006). *Moçambique – 2006–09 Contry Strategy Paper.* Country Operations Department.

All Africa (2011). *Moçambique: Government Approves Mphanda Nkuwa Dam,* allafrica.com/stories/201008190343.html (accessed 1 Sep 2008).

Arndt, C. (2007). *Comments on : "The war ended 15 years ago but we are still poor" authored by Joseph Hanlon.* Cape Town: National Directorate of Studies and Policy Analysis, Ministry of Planning and Development.

Askouri, A. (2007). China's investment in Sudan: displacing villages and destroying. In F. Manji, and S. Marks (eds), *African perspectives on China in Africa.* Nairobi: Fahamu.

Brown, O., Hammil, A., & Maclemman, R. (2007). Climate change as the 'new' security threat: implications for Africa. *International Affairs* 6 (83).

Bucuane, A., & Mulder, P. (2007). *Exploring and Electricity Tax on Mega Projects in Moçambique.*

Maputo: National Directorate of Studies and Policy Analysis, Ministry of Planning and Development.

Burke, C. (2007). Evaluating China's involvement in Sudan's Merowe Dam Project. *China Monitor* 7 (9).

Business Map. (2000). *SADC Investor Survey: complex terrain.* Johannesburg: Business Map.

Cox, R. (1996). Global Perestroika. In R. Cox, *Approaches to World Order,* p. 296–313. Cambridge: Cambridge University Press.

Cuereneia, A. (2001). *The process of decentralisation and local governance in Moçambique: experiences and lessons learnt.* Cape Town: United Nations Capital Development Fund.

Engineering News (2008). *Moçambique could help South Africa's energy crisis.* Accessed June 29, 2008, from Engineering news, <www.engineeringnews.co.za/article.php?a_id=125539>.

Eriken, S., O'Brien, K., & Rosentrater, L. (2007). *Climate change in eastern and southern Africa: impacts, vulnerability and adaptation.* Oslo: Global Environmental Change and Human Security.

Ghazvinian, A. (2008). *Impacts of Deforestation in Moçambique.* Accessed June 15, 2008, from Village Reach, <www.villagereach.org/PDF-Documents/Impacts-of-Deforestation-in-Mozambique.pdf>.

International Rivers Network. (2006). *Mega–Barragem Moçambicana vai agravar a pobreza e a subsistência.* <www.internationalrivers.org/resources> (accessed 20 Feb 2008).

Jornal Notícias de Moçambique (2007 A). Cooperação com a China: PM enaltece ausência de condicionalismos. *Jornal Notícias de Moçambique.*

Jornal Notícias de Moçambique. (2007 B). Tecnologias agrícolas e infra-estruturas: China disponibiliza fundos. *Jornal Notícias de Moçambique.*

Justiça Ambiental and IRN. (2006 B). *Damning the Zambeze – risks outweight the benefits of proposed Mphanda Nkuwa Dam.* International Rivers Network Publication.

Lemos, A., and Ribeiro, D. (2007). Taking ownership or just changing owners? In F. Manji, and S. Marks (eds), *African Perspectives on China in Africa.* Fahamu and Pambazuka.

Lusa (2008, March 24). *Brasil: Construção de barragem em Moçambique é "prova de fogo" para empresa brasileira.* Accessed April 20, 2008, from Radio Televisão Portuguesa, ww1.rtp.pt/noticias/index.php?article=335173&visual=26&rss=0.

Meguigy N. (2002). *Mphanda Uncua Hydropower Project Mozambique.* Technical Unit for the Implement of Hydropower Projects.

Morrissey, J. (2006). *Meios de Subsistência e Risco: O caso da barragem de Mphanda Nkuwa.* Justiça Ambiental.

Portuguese News Network (2011). Moçambique: Justiça Ambiental alerta para impactos negativos da barragem Mphanda Nkuwa, <www.jornaldigital.com/noticias.php?noticia=25682>.

Reuters (2010). Mozambique approves $2 bln hydroelectric dam-paper: <af.reuters.com/article/topNews/idAFJOE67H05L20100818>.

Soderbaum, F., & Taylor, I. (2001). Transmission belt for transnational capital or facilitator for development? Problematising the role of the state in the Maputo Development Corridor. *Journal of Modern African Studies*, 39 (4): 675–695.

Soren Ambrose, L. B. (2008, April 1). *Leaky logic: dams in three countries questioned.* Accessed June 15, 2008, from Bretton Woods Project, <www.brettonwoodsproject.org/art-561036>.

Yeld, J. (2008, March 6). *Eskom may call for two time zones to spread demand.* Accessed April 11, 2008, from DST News, <www.worldtimezone.com/dst_news/dst_news_southafrica02.html>.

Yi-Chong, X. (2004). *Electricity Reform in China, India and Russia.* Cheltenham: Edward Elgar Publishing.

PART III
Water supply and sanitation

The water question under extreme poverty. The example of Bolama, Guinea-Bissau (West Africa)

The politics of water management in rural communities of Nigeria

Adriano Bordalo[*] and Joana Savva-Bordalo[**]

The water question under extreme poverty: The example of Bolama, Guinea-Bissau (West Africa)

Abstract. The lack of access to safe water is a burden that affects millions worldwide. In spite of progress in sanitation, water supply and wastewater treatment, the reappearance of waterborne diseases associated with poverty and climate change clearly show the need for a new paradigm in dealing with the water question, particularly in Africa. Guinea-Bissau (West Africa) is one of the least developed countries in the world. Almost 90% of the population has no access to tap water, obtaining the water for daily needs essentially from wells and drilling wells without any quality screening. The water vulnerability may be even higher during the dry season (that lasts 6 months) when several wells dry and the population, mostly women and children, must walk long distances to search for water. This work was carried out in Bolama (11.5°N; 15.4°W), the island capital of the Bijagós Archipelago, and aimed to better understand the seasonal variability of domestic water quality, the prevalence of disease that affects the population and the design of low-cost measures for improving water quality in the absence of a large-scale plan to provide safe water and sanitation under extreme poverty. The results are essential for the development and implementation of sustainable measures in order to minimize the risks associated with water of doubtful quality, the only one available to the public.
Keywords. Bolama, Guinea-Bissau, well, water quality, disease.

* Corresponding author. Institute of Biomedical Sciences (ICBAS), University of Porto (UP), Lg. Abel Salazar 2, 4099–003 Porto, Portugal. Ciimar-Centre of Marine and Environmental Research, Rua dos Bragas, 289, 4050–123 Porto, Portugal. bordalo@icbas.up.pt.

** Instituto Português de Oncologia, Rua Dr. António Bernardino de Almeida, 4200–072 Porto, Portugal.

Introduction

Water is a source of life and a natural resource that sustains our environments and supports livelihoods (UNDP, 2006), but can also pose a health risk due to its scarcity and/or lack of quality. Millions of people around the world remain without access to potable water supplies and lack adequate methods for disposing of human fecal waste (Esrey and Habicht, 1986). It is estimated that two in five people worldwide do not have access to improved sanitation (UNICEF, 2006). This recurrent situation threatens human development and particularly the future of the most vulnerable segment of the population: children. Over 9% of the global disease burden could be prevented by introducing better management practices (Bartram, 2008). Moreover, due to the lack of safe water and in spite of the improvements, recent data show that 1.8 million people die every year from diarrheal diseases, the vast majority children under 5 (WHO, 2007). Diarrhoea is the leading cause of morbidity and mortality in children in the developing world (Bern et al., 1992). The importance of sanitation in preventing diarrheal diseases was recognized in the Millennium Development Goals (UN, 2006), which set a target of halving the number of people without access to basic sanitation by 2015. Moreover, improvements in drinking water, sanitation facilities, and hygiene practices may pave the way for the reduction of illness in lesser developed countries (Fewtrell, 2005), easing the burden of severe human suffering and economic loss.

In addition to diarrhoea, other waterborne and water related/dependent diseases plague the world, particularly the less developed countries. The effect of tuberculosis or HIV on humans can be greatly amplified due to poor hygienic conditions as a result of the lack of potable water and proper sanitation. Those diseases are a serious calamity, especially in Africa, where prevalence can reach 1,137 per 100 000 people for the former and 33.4% for the latter, in Djibouti and Swaziland respectively (UNDP, 2006). Moreover, malaria is another cruel example of a water-related vector disease that kills more than two million people worldwide, >90% of them in Africa, half of which children (Martens and Hall, 2000). High malaria incidence has been related to poor water quality (Srivastava et al., 1999), although malaria vectors can be controlled through different water management practices (Konradsen et al., 1998).

The water vulnerability that affects billions worldwide may be further exacerbated by the current climate change, with Africa constituting the most vulnerable continent. Sub-Saharan Africa lies in a region where rainfall is seasonal and at least during half the year arid conditions predominate, affecting crops that are the basis of the diet (e.g. rice, cassava). The risk of crop failure is high, a fact which makes domestic food supplies insecure (Falkenmark, 1990). This, in turn, has a negative impact on humans and is worsened by the lack of proper water supplies and sanitation. However, little is known about the seasonal variability of water quality, particularly in areas where piped water is not available and the population relies on wells, streams and rivers as the source of water for all domestic needs. In addition, some 2.6 billion people spend an important proportion of their time in simply securing drinking water or a place to defecate (WHO/UNICEF, 2006).

In order to contribute to filling in the gap regarding water quality dynamics and their impact on human health in Sub-Saharan Africa, this paper addresses (i) the seasonal variability of domestic water quality on Bolama Island (Guinea-Bissau), one of the 10 poorest countries in the world, where hand-dug wells are the sole source of water for the population; (ii) the prevalence of disease affecting the islanders; and (iii) identifying low-cost, easily applicable measures for improving water quality in the absence of a strategic plan to supply safe water and sanitation.

Material and methods

The environment

The Republic of Guinea-Bissau (11° N; 15° W) is a coastal country located in sub-Saharan West Africa (Figure 1). Most of its 28.000 km^2 land area is covered with tropical rain forest (85%). Two well-marked seasons exist:

wet with SW winds from June to November and dry, with NE winds from December to May.

Bolama Island (100 km^2) is the provincial capital and the most densely inhabited island of the Bijagós Archipelago off the coast of Guinea-Bissau. The hydromorphic sandy-clay soils are acidic, with low organic matter contents. The sediments are mainly Cenozoic to Recent (Teixeira, 1968). The vegetation consists of dry, deciduous forest and semi-deciduous rain

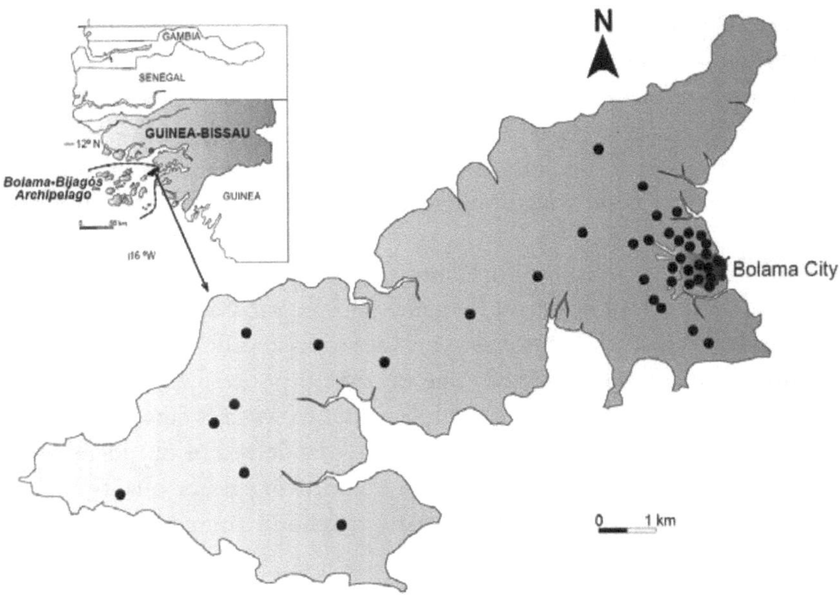

Figure 1. Bolama Island (Guinea-Bissau) and location of sampling stations.

forest, palm and chew-nut forests, coastal and wet savannas and extensive mangrove forests lining the coastal one-third of the island.

All ethnic groups are represented and the life expectancy at birth of the 6,500 inhabitants is low (46.9 years). Health infrastructure is minimal and was greatly affected during the civil unrest in the late 1990s. Communicable diseases such as endemic malaria, diarrhoea (including cholera) and respiratory infections predominate. Each person has only

21 litres of water available for daily personal needs (Bordalo and Savva-Bordalo, 2007). No continuous electricity is supplied and no working water or sanitation networks are available. Sanitary wastewater is disposed off in public and private latrines as well as in the fields, often close to the wells that supply the population with water.

Sampling strategy

Two surveys were conducted. The first in July 2006 corresponded to the wet-season (WS), whereas the dry-season (DS) survey was performed in February 2007. Between 28 (WS) and 34 wells (DS) were surveyed, taking into account accessibility, pump functionality (in the case of wells equipped with pumps) and the presence of water. The exact position of each well was obtained by means of GPS (Magellan 600). At each site digital photographs were taken in order to document the surrounding environment.

Water was collected by pumping or by means of a bucket and the initial 10 litres (one bucket) were discarded. A 500 ml plastic sterile flask was filled as well as one 1.5 l clean plastic bottle. In the latter case, bottles were washed and rinsed with 500 ml of the sample before filling. Immediately after collection, the samples were refrigerated with locally produced shredded ice in 48 l chests and transported to the field laboratory for analysis no later than 3.5 hours after collection.

Physical and chemical analysis

In situ measurements of temperature, conductivity, dissolved oxygen, oxygen saturation were performed either with a YSI 6120 multiprobe (WS) or with a Hanna Instruments HI9828 multiprobe (DS), calibrated according to the manufacturers' specifications. Turbidity (Tur) was also measured during the WS with the same probe, whereas in the DS it was assayed photometrically with a Hanna Instruments C114 meter. Additional chemical parameters were assayed in the field laboratory with a Hanna Instruments C200 multiparameter ion specific meter, according to the standard methods suggested by the manufacturer. Arsenic concentrations were measured with a high sensitivity Merckoquant As kit (ref 17927) (Table 1). Biological oxygen demand was measured as the difference in oxygen concentrations before and after five days incubation at 20–22°C. Incubation was performed in an ice chest in the dark and temperature regulated by periodically adding shredded ice. Chemical oxygen demand

Parameter	Unit	Method
Temperature	°C	Electrode
Conductivity	$\mu S\ cm^{-1}$	Electrode
Dissolved oxygen	mg/L	Electrode
pH	-	Electrode
Turbidity	NTU	Electrode
Colour	PtCo	Colorimetric
Nitrate mg	$NO3\ l^{-1}$	Cadmium reduction
Ammonium	$mg\ NH4\ l^{-1}$	Nessler
Nitrite	$mg\ NO2\ l^{-1}$	Diazotization
Copper µg	$Cu\ l^{-1}$	Bicinchoninate
Iron	$\mu g\ Fe\ l^{-1}$	TPTZ
Chromium	$\mu g\ Cr\ l^{-1}$	Diphenylcarbohydrazide
Cyanide	$\mu g\ Cn\ l^{-1}$	Pyridine-pyrazolone
Arsenic	$\mu g\ As\ l^{-1}$	Mercury bromide
Aluminum	$\mu g\ Al\ l^{-1}$	Aluminon

Table 1. Parameters and methods for physical and chemical analysis.

was assayed photometrically after acid-digestion in a dry bath for 2 h at 150°C (Merck Spectroquant COD cell test). All methods were previously checked in the main laboratory (Portugal) for accuracy against standards and are reported elsewhere (Bordalo and Savva-Bordalo, 2007).

Microbiological analysis

Each water sample was filtered through a sterile disposable filtering devise (Pall Co) onto a gridded 0.45 μm membrane filter provided with the individual kit. Vacuum was obtained by means of a hand pump (Mityvac). Asepsis was achieved with an alcohol burner, and the working bench was covered with a thick plastic sheet periodically disinfected with absolute alcohol. Duplicate filters were placed in 60 mm Petri dishes containing mFC agar (Difco) for fecal coliforms (FC) and Slanetz-Bartley agar (Oxoid) for fecal enterococci (FE). Incubations were performed in mini-incubators (Merck) at 44.5°C for 24 h. Typical colonies were counted and results expressed as colony forming units (CFU) 100/mL.

Water quality index

The water quality index (WQI) used is a tropical modification of the original Scottish Water Quality Index (Bordalo et al., 2001) and has been previously applied to Bolama Island water conditions (Bordalo and Savva-Bordalo, 2007). The 10-parameter index consists of the aggregation of three groups of parameters: physical (temperature, conductivity, suspended solids), chemical (pH, ammonia, nitrate, dissolved oxygen), and organic/microbiological (fecal coliforms, biological oxygen demand, chemical oxygen demand). The chosen method for subindices aggregation (weighted arithmetic average) is particularly suitable for indexation of the general water quality, as stated by House (1989). The final, modified, arithmetic, weighted index is the result of squaring the sum of the products of water quality ratings (qi) and weighting each individual parameter (wi) divided by 100, according to the following equation:

$$WQI = 1/100\left(\sum_{i=1}^{9} q_i w_i\right)^2$$

In the SDD WQI, 0% represents the poorest and 100% the highest water quality. In this study, the House and Ellis (1987) class rating was adopted: 10–25%, badly polluted; 26–50%, polluted; 51– 70%, reasonable; 71–90%, good; 91–100%, very good.

pH experiments

The acidity assay was conducted in 15-l water buckets similar to those used by the population to retrieve and store water indoor. Exactly 10-l of well water were measured and 12 oyster shells (400 g) were added to the bottom of the bucket. The empty shells were collected on the beach, washed, rinsed with distilled water, baked under sunlight for 1 d and rinsed again with distilled water. The above-mentioned HI9828 multiprobe instrument was immerged in the bucket and data gathered every 5 min for 12 consecutive hours. Three experiments were conducted with initial pHs of 4.67, 5.12 and 5.60 at room temperature (24–29°C). Due to the existence of a single measurement device, no replicates were performed.

Health and meteorological data

The health data presented in this paper span over a 24-month period from January 2006 to December 2007, covering entirely the 2006 WS and the 2006/2007 DS. Data were obtained from the Bolama Regional Hospital logbook. Meteorological data were obtained from the Bolama Regional Meteorological Service (precipitation) and from the Guinea-Bissau Meteorological Service in Bissau (temperature).

Data treatment

Bacterial data were log (n+1) transformed prior to analysis. Health data were seasonally aggregated as WS and DS. Furthermore, data were tested for normality, using the Kolmogorov-Smirnov test, and for homoscedasticity using Leven's test (Zar, 1996). Least-square regression was used to compare relationships between two variables. Mapping was performed with Surfer using krigging as the interpolating method. In the absence of Guinea-Bissau water quality standards, the European (EU, 1998) and WHO recommendations (WHO, 2006) were adopted.

Results

Environmental characteristics

Air temperature and rainfall for the period 2006–2007 are plotted in Figure 2. Temperatures during the DS are higher than in the WS, with May being the warmest month (27.95 °C and 27.70 °C, respectively) and January the coolest (25.3 °C and 25.7 °C, respectively). During the above-mentioned period, total precipitation was 1,768 mm for 2006 and 1,751 mm in 2007. No significant ($p > 0.05$) differences were found. August was the wettest month with 527 mm and 499 mm, respectively. In August, the number of wet days reached 22 in both years.

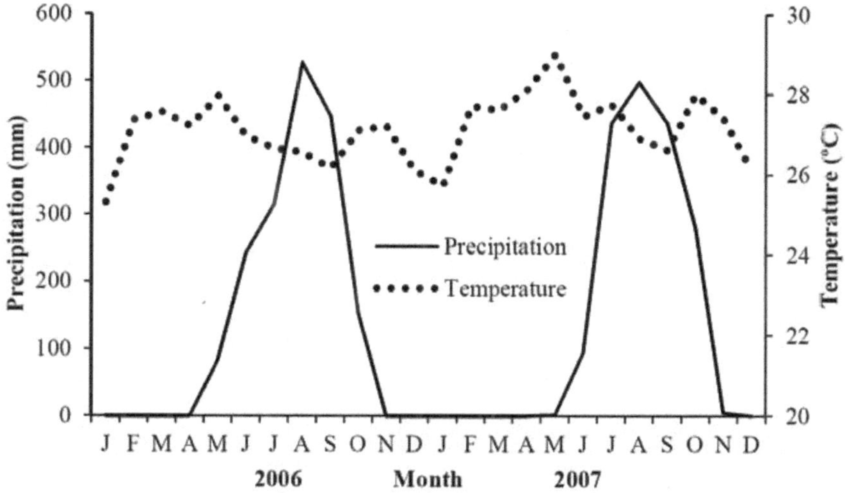

Figure 2. Monthly temperature and precipitation for the period 2006–2007 on Bolama Island.

As expected, the distance from the land surface to the top of the water was less in the WS due to natural recharge from rainfall that fostered the increase of the water volume in the shallow Bolama wells. Physical and chemical water quality parameters such as temperature, conductivity, dissolved oxygen, turbidity, as well as nitrite, copper and cyanide showed

Parameter	Unit	Wet Season			Dry Season		
		Average	Median	SE	Average	Median	SE
Depth	M	9	7	1	12	12	1
Temperature	°C	29.05	29.29	0.15	28.68	29.00	0.27
Conductivity	µS cm^{-1}	136	107	18	132	100	17
Oxygen saturation	%	87	91	3	48	48	2
Dissolved oxygen	mg l^{-1}	6.58	6.95	0.24	3.64	3.57	0.15
pH	-	5.12	5.22	0.08	4.98	5.15	0.09
Turbidity	Ntu	6.5	3.0	1.5	3.3	0.1	2.7
Colour	PtCo	31	7	9	40	29	11
Nitrate	mg l^{-1}	16.60	14.83	2.44	22.58	20.38	2.93
Nitrite	mg l^{-1}	0.04	0.03	0.00	0.03	0.00	0.01
Ammonium	mg l^{-1}	0.11	0.01	0.05	0.17	0.10	0.05
Copper	µg l^{-1}	95.5	49.0	19.2	67.9	31.0	21.6
Iron	µg l^{-1}	66.5	21.5	18.7	213.4	107.0	82.8
Chromium	µg l^{-1}	26.5	25.0	2.9	10.2	2.0	4.9
Cyanide	µg l^{-1}	12.1	10.0	2.0	5.6	2.0	1.6
Arsenic	µg l^{-1}	1.0	1.0	0.2	1.4	0.0	0.3
Aluminum	µg l^{-1}	86	60	26	300	210	87
Fecal coliforms	cfu 100 ml^{-1}	410	35	201	16,008	103	11,422
Fecal enterococci	cfu 100 ml^{-1}	74	3	35	186	4	147

Table 2. Average, median and standard error (SE)
for physical, chemical and on Bolama Island, Guinea-Bissau.

higher values during the WS, whereas color, nitrate, ammonium, iron, chromium arsenic and aluminum increased during the DS. All samples showed a pH lower than the EU parametric value of 6.5. Water acidity increased slightly during the DS (4.98±0.09), but the lowest pH (3.48) was found during the WS in one unprotected well on the SW tip of the island that serves 500 people. Also, iron, chromium and aluminum contents nearly triplicated during the DS, probably due to the soil characteristics and lower dilution.

As far as microbial characteristics are concerned, fecal contamination was significantly higher ($p<0.05$) in the DS versus the WS. The absence of rain and continual contamination of the wells were the main responsible factors. On average, fecal coliforms were two orders of magnitude higher whereas fecal enterococci increased by one order of magnitude (Table 2). Regardless of the season, eight out of ten wells showed fecal contamination levels that threaten the safety of water consumption without treatment.

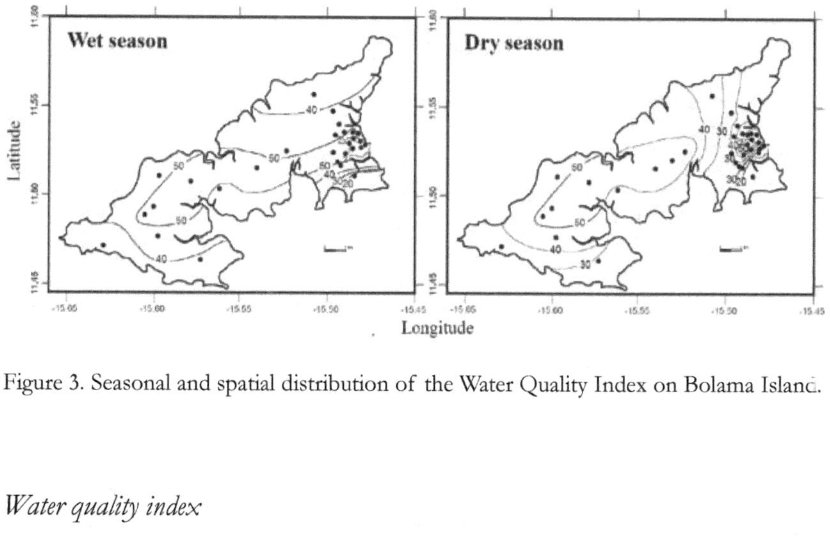

Figure 3. Seasonal and spatial distribution of the Water Quality Index on Bolama Island.

Water quality index

The spatial distribution of seasonal water quality is plotted in Figure 3. The average WQI was 43±4% during the WS, a rather low value even for the production of potable water. The WQI further decreased in the DS to 32±3%. During the former season, factors such as high acidity and fecal contents were responsible for the lower score, whereas during the latter, in addition to the forementioned parameters, low oxygen contents and high organic levels from BOD_5 measurements contributed to the decrease in quality.

Spatially, the trend was very similar: water quality increased in the central parts of the island but decreased in the urban area as well as on the SW tip of the island. However, some of the wells that were free of fecal contamination during the WS were found to be slightly polluted in the DS requiring additional attention.

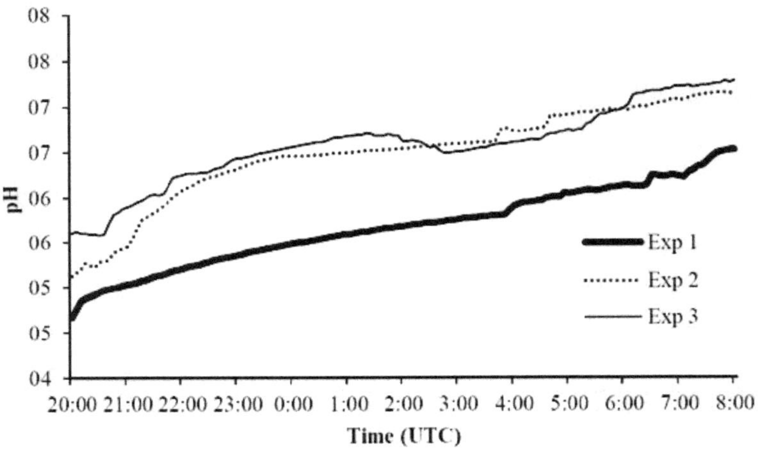

Figure 4. Time series of pH variability in water amended with oyster shells.

pH experiments

Regardless of the initial pH (4.48 and 5.39), the overnight 12-h incubation period with oyster shells decreased the acidity to pH 6.64 and pH 7.04, respectively. Figure 4 presents the trend for the second experiment. It should be noted that conductivity also increased monotonically but to values <600 μS cm^{-1}, as in the case of the more acidic experiment. The presence of oyster shells did not alter the colour or the taste of the water.

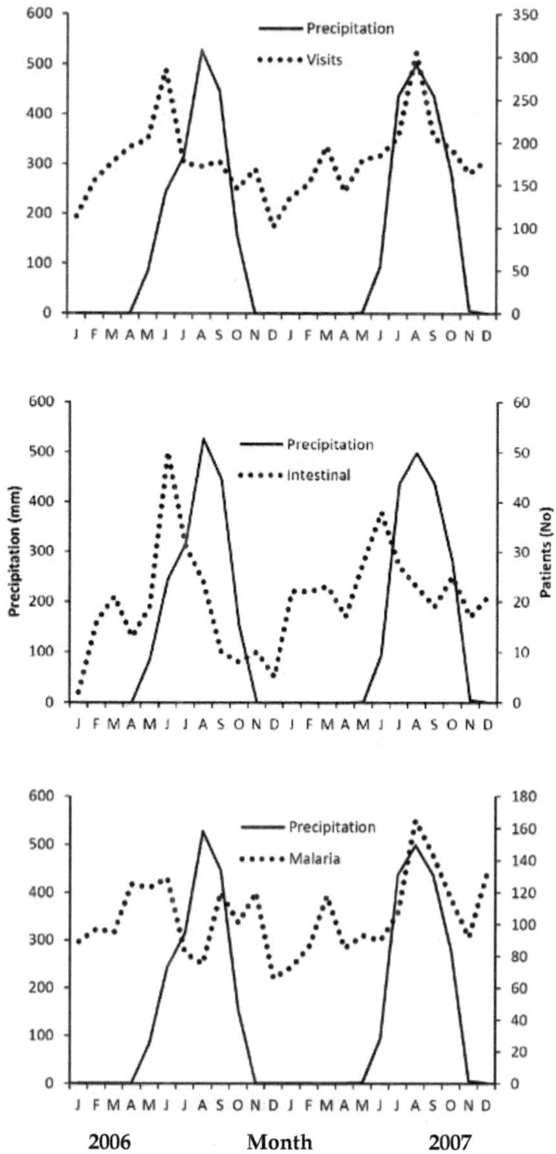

Figure 5. Monthly variability of precipitation, total number of healthcare hospital visits, number of patients with gastrointestinal diseases, and number of malaria cases.

Human health

From January to December 2006, a total of 2,070 patients, 55.7% of which were children < 15 years old, visited the Bolama regional hospital and were registered in the logbook. In the same period of 2007, a total of 2,247 visited the hospital (62.8% children). Females represented more than a half of the patients (50.9% in 2006, 54.1% in 2007).

In 2006, at the onset of the WS (June), the number of hospital visits increased dramatically, whereas in 2007 the rise occurred in August, when precipitation peaked (Figure 5). Gastrointestinal diseases (diarrhoea, gastroenteritis, intestinal parasites) always peaked in June and affected 7.6% and 12.6% of the total number of patients in 2006 and 2007, respectively. Children were particularly vulnerable representing two out of three patients. Male children demonstrated more gastrointestinal disturbances (52.1%) than female children. In the case of adults, a reverse situation was found, with females representing 55.3% (2006) or 66.7% (2007) of the total.

The total incidence of malaria was 57.8% in both periods, being 50.7% (2006) and 55.4% (2007) for adults and 63.4% (2006) and 59.2% (2007) for children. No gender differences appeared for this disease in children, whilst a totally different situation emerged regarding adults. Females with malaria outnumbered males by 1.5x representing 66.2% (2006) and 67.0% (2007) of cases.

Discussion

In Africa, as in many areas of the developing world, population growth is resource determined, with the lack of potable water being one of the major development constraints. Even in the case of access to water in quantity, for example from rivers, wells or networks, the lack of quality may impair the life of millions of humans that must struggle desperately for survival. The calamitous vulnerability regarding water sources may be compounded by the current climate change in Africa, the most vulnerable continent.

Sub-Saharan West Africa is located in a zone where rainfall is seasonal and during at least half the year arid conditions predominate affecting those crops that are the basis of the diet (e.g. rice, millet, cassava). Thus, the risk of crop failure is high making domestic food supplies insecure (Falkenmark, 1990). The negative impact on humans can be further magnified by the lack of proper water supplies. Actually, the recurrent anomalous wet or dry conditions, may affect the influence of water quality on human health. Diarrhoea, a water born disease, is the leading cause of morbidity and mortality of children under five years old in developing countries (Bern et al., 1992) and the latest outbreaks of cholera in west Africa (Congo-Angola in 2006, Guinea-Senegal in 2006, Guinea-Bissau in 2008 and Zimbabwe in 2009) which have been linked to particularly wet conditions are a good example.

On Bolama Island, all water used for domestic consumption and irrigation is retrieved from shallow wells, which, in most cases, are within short distance of the houses. This facilitates the collection of the water from the wells. Gender disparity exists regarding water collection and transportation. During the morning the women bear the main responsibility for water retrieval, whereas during the rest of the day children regardless of sex perform the task. It is estimated that in Bolama the per capita consumption of water for drinking, cooking and bathing doubles the national Guinea-Bissau average of 21 l (Bordalo and Savva-Bordalo, 2007). After use, water is disposed of within close vicinity to the houses and quickly infiltrates through the soil. Usually, latrines do not have water for hand washing or for cleaning the premises and are roof-less. Domestic animals are not fenced and wander freely, often within vicinity of the wells, as in the case of pigs. They are not fed, and human excreta are part of their meals.

The quality of drinking water is of growing concern worldwide because of microbial contamination and hazardous levels of chemicals such as arsenic and fluoride (WHO/UNICEF, 2006). In the case of Bolama, no chemical pollution was generally detected, except for iron and aluminum in selected wells during the DS, probably due to the absence of industry and the limited use of fertilizers and pesticides in the fields. However, due to the high levels of fecal indicator bacteria and water acidity detected, the use of wells as the only source of water poses a serious health hazard, a situation not always recognised by users due to the lack of proper information. A strong seasonal water quality signal was found. Overall quality decreased dramatically during the 2007 DS. The lack of rainfall and the shallowness of the wells prevented the renewal of the water by oxygen-rich rain water from the surface. Owing to the acidic nature of the sandy-clay soils characterising the island, the underground water that feeds the wells has a low pH. Acidity levels increased during the dry period of the year probably due to higher residence of water inside the wells. Concomitantly, conductivity also increased. Since no sewage network exists, defecation occurs in latrines, which are of limited use in many villages, or directly into the fields, often close to wells that supply the water for domestic consumption. Surprisingly, fecal contamination was lower in the WS probably due to higher dilution in clear contradiction with disease indicators (see below). Spatially, fecal contamination decreased in wells within the central region of Bolama Island where latrines are kept at a distance >30 m, although this trend was more evident during the WS. Independently of the season, four out of fives wells were contaminated with fecal materials which meant that the population that has no alternative but to consume polluted water.

The water quality index used in this work was intended to provide a simple and understandable tool for managers and decision makers regarding the quality and possible uses of a given body of water (Bordalo et al., 2001). Basically, this index attempts to provide a method for presenting a cumulatively derived, numerical expression that can define the level of water quality (Miller et al., 1986). Fecal contamination and low pH have been previously shown (Bordalo and Savva-Bordalo, 2007) to be the main factors responsible for the low WQI scores, regardless of the season. During the DS, low oxygen contents and high organic matter levels had a synergetic effect for the obtained lower scores. Higher WQI values were found in the central part of the island, where wells are separated from latrines and bathing areas by at least 30 m, and animals are fenced. Using the rating scale proposed by House and Ellis (1987), we found that even in the few cases where the WQI was >50%, i.e., falling within the reasonable category, there was an imperative requirement for proper treatment to increase potability, which is neither performed on Bolama nor in Guinea-Bissau.

Although the fecal contamination of well water was higher in the DS than in the WS, the number of patients with gastrointestinal diseases, diarrhoea included, was higher in the wet period of the year, particularly at the onset of the WS. This shows that, besides the risk associated with the continual use of highly contaminated water for drinking, cooking and bathing, surface stagnant water that attracts insects and domestic animals that are in contact with humans, may emerge as an additional and major factor contributing to the degradation of health. Our results clearly show that adult females are more prone to gastrointestinal disease than adult males. In reality, females are in closer contact to water. Not only are they are responsible for collecting water by carrying on their heads up to 40 liters on each of multiple trips, but they also do the laundry and dishwashing, very often within close range of the wells.

Malaria is a water related disease that affects the entire country, accounting for 57.8% of all healthcare visits. This is in line with the Guinea-Bissau average of 58% (Lis, 2003). In 2007, the number of patients peaked during the wettest month (August) and was clearly offset by the gastrointestinal increase in June. These seasonal dynamics for the two locally predominant illnesses should be taken into consideration in future plans to control disease in Bolama and Guinea-Bissau, particularly in the field of human recourses, health care facilities and medical supplies, including appropriate medications.

The morbidity and mortality rate of water related diseases could be reduced dramatically, through the provision of safe drinking water and adequate sanitation (Helmer, 1999). Although it is acknowledged that home water treatment does not replace a drinking-water infrastructure, in the absence of plans for the rehabilitation of the existing but nonfunctional water network or to improve the access to safe water for the entire population, proactive and sustainable measures should be designed. In fact, amongst people who lack access to safe water, point-of-use water treatment may be one of the most efficacious strategies for reducing diarrhoea (Fewtrell et al., 2005), a disease that kills annually 1.8 million worldwide (WHO, 2007). A broadly disseminated and tested method for improving the microbiological quality of drinking water at the household level is the use of diluted chlorine solution (Arnold and Colford, 2007). On the other hand, chlorination with bleach is an effective method for decreasing the pathogenic risk of drinking water (Mintz el al., 1995), although it is less efficient in inactivating viruses and is of limited effectiveness against parasites (Mintz et al., 2001). Liquid bleach is a household product locally available at a rather moderate price. A simple experiment performed on location showed that a single drop of bleach in 1 liter of well water was able to inactivate fecal coliforms within 60 min of contact (data not shown). This is in line with previously reported findings (e.g. Quick et al., 1996; Wright et al., 2003; Arnold and Colford, 2007). However, chlorination is not a widespread practice in Guinea-Bissau, except during cholera outbreaks when a much criticized WHO practice of single shock chlorination of wells with bleach (WHO, 2005) instead of point-of-use is performed (Rowe et al., 1998). The absence of information by the competent authorities regarding the benefits of such simple procedures as point-of-use water treatment to decrease the pathogenic risk mainly associated to bacteria through the oral-fecal disease transmission route, has led to the present situation in Guinea-Bissau in parallel with most Sub-Saharan Africa, where central infrastructure for treating drinking water do not exist. Moreover, that part of the world remains the only region where, over the period 1990–2004, the number of people without access to safe drinking water increased by 23% (WHO/UNICEF, 2006). Another point of infection deserves attention – recontamination of stored drinking water by users and domestic animals. In Bolama the entire family drinks from 15–25 L wide mouth earthen jar by dipping the same handled metal cup into the water without washing hands or the cup itself, a procedure that leads to the deterioration of the home stored water (Pinfold, 1990). Moreover, such post-source contamination has been shown to increase diarrhoea risk (Clasen et al., 2006), highlighting the need for improved personal and domestic hygiene practices (Oswald et al., 2007). The use of safe storage vessels like narrow-mouthed closed containers fitted with a spigot (Witt and Reiff, 1993) could reduce the risk of recontamination associated with water storage and retrieval and was recommended (but never implemented) in Guinea-Bissau during the cholera outbreak of 1996–1997 (Daniels et al., 1999).

Another single factor that impairs the water quality in Bolama is the low water pH, well below the recommended threshold of 6.5 by European standards. It was found that the overnight immersion of 400 g of calcium carbonate rich oyster shells, a freely available mollusk in Guinea-Bissau mangrove belt that lines the coastal zone, in a 10-l water bucket was enough to increase pH to values ≥ 6.5, reducing the corrosive power associated to low pH (Figure 4). Moreover, the neutralized water is also a potential source of calcium, a much-needed, under-consumed nutrient in the Guinea-Bissau diet.

It is unlikely that the urban Bolama water network will be rehabilitated in the near future despite the existence of a World Bank plan (World Bank, 1997). Thus, the access to freshwater for the entire population will remain restricted to wells with doubtful water quality, as the sole water source for all domestic uses as well as for crop irrigation. The proactive measures presented in this study, based on the evaluation of the seasonal water quality and on health statistics from the Bolama Regional Hospital, were designed in a sustainable fashion in a country suffering from chronic poverty, where good health remains elusive for a large fraction of the population. However, this alone is not enough. Proper management plans should be drawn up by the central and regional governments, with the involvement of the public, targeting the promotion of health and reduction of poverty, with the potential participation of international donors and experts with field experience.

Acknowledgments

This work was partially sponsored by *Assistência Médica Internacional (AMI)*, a Portuguese NGO with a permanent base in Bolama, by providing a 4-wheeel vehicle on Bolama Island and boat transportation to and from the mainland (Bissau city).

The work was partially funded by FCT grant PTDC/AAC-CLI/ 103539/2008.

References

Arnold, B. F. and Colford Jr., J. M. (2007). Treating water with chlorine at point-of-use to improve water quality and reduce child diarrhea in developing countries: a systematic review and meta-analysis. *American Journal of Tropical Medicine and Hygiene* 76: 354–364.

Bartram, J. (2008). Flowing away: water and health opportunities. *World Health Organization Bulletin* 86: 1–2.

Bern, C., Martines, J., de Soysa, I. and Glass, R. I. (1992). The magnitude of the global problem of diarrhoeal disease: a ten-year update. *World Health Organization Bulletin* 70: 705–714.

Bordalo, A. A. and Savva-Bordalo, J. (2007). The quest for safe drinking water: an example from Guinea-Bissau (West Africa). *Water Research* 41: 2978–2986.

Bordalo, A. A., Nilsumranchit, W. and Chalermwat, K. (2001). Water quality and uses of the Bangpakong River (Eastern Thailand). *Water Research* 15: 3635–3642.

Clasen, T., Roberts I., Rabie T., Schmidt W. and Cairncross S. (2006). Interventions to improve water quality for preventing diarrhoea. *Cochrane Database Systematic Reviews* 19: CD004794.

Daniels, N., Simons, L., Rodrigues, A., Gunnlaugsson, G., Forester, T., Wells, J,. Hutwagner, L., Tauxe, R. and Mintz, E. (1999). First do no harm: making oral rehydration solution (ORS) safer in a cholera epidemic. *American Journal of Tropical Medicine and Hygiene* 60: 1051–1055.

Esrey, S. A. and Habicht, J. P. (1986). Epidemiology evidence for health benefits from improved water and sanitation in developing countries. *Epidemiologic Review* 8: 117–128.

EU (1998). *Council Directive 98/83/EC of 3 November 1998 on the quality of water intended for human consumption*, L 330, 05/12/1998, 32–54.

Falkenmark, M. (1990). Rapid population growth and water scarcity: the predicament of tomorrows Africa. *Population and Development Review* 16: 81–94.

Fewtrell, L., Kaufmann, R. B., Kay, D., Enanoria, W., Haller, L. and Colford Jr., J. M. (2005). Water, sanitation, and hygiene interventions to reduce diarrhoea in less developed countries: a systematic review and meta-analysis. *The Lancet Infectious Diseases* 5: 42–52.

Helmer, R. (1999). Water quality and health. *The Environmentalist* 19: 11–16.

House M. A. and Ellis J. B. (1987). The development of water quality indices for operational management. *Water Science and Technology* 19: 145–154.

House, M. A. (1989). A water quality index for river management. *Journal, Institution of Water Engineers & Scientists* 3: 336–344.

Konradsen, F., Matsunoa, Y., Amerasingheb, F. P., Amerasingheb, P. H and van der Hoeka, W. (1998). 'Anopheles culicifacies' breeding in Sri Lanka and options for control through water management. *Acta Tropica* 71: 131–138.

Lis, J. (2003). *Guinea-Bissau health assessment.* LEAD International Ministry Network.

Martens, P. and Hall, L. (2000). Malaria on the move: Human population movement and malaria transmission. *Emerging Infectious Diseases* 6: 103–109.

Miller, W. W., Joung, H. M., Mahannah, C. N. and Garrett, J. R. (1986). Identification of water quality differences in Nevada through index application. *Journal of Environmental Quality* 15, 265–272.

Mintz, E. D., Reiff, F. M. and Tauxe, R. V. (1995). Safe water treatment and storage in the home: a practical new strategy to prevent waterborne disease. *Journal of the American Medical Association* 273: 948–953.

Mintz, E., Bartram, J., Lochery, P. and Wegelin, M. (2001). Not just a drop in the bucket: Expanding access to point-of-use water treatment systems. *American Journal of Public Health* 91: 1565–1570.

Oswald, W. E., Lescano, A. G., Bern, C., Calderon, M. M., Cabrera, L. and Gilman, R. H. (2007). Fecal Contamination of Drinking Water within Peri-Urban Households, Lima, Peru. *American Journal of Tropical Medicine and Hygiene* 77: 699–704.

Pinfold, J. V. (1990). Faecal contamination of water and fingertiprinses as a method for evaluating the effect of low-cost water upply and sanitation activities on faeco–oral disease transmission. I. A case study in rural north-east Thailand. *Epidemiology and Infection* 105: 363–375.

Quick, R., Venczel, L., Gonzalez, O., Mintz, E., Highsmith, A., Espada, A., Damiani, E., Bean, N., De Hannover, R. and Tauxe, R. (1996), Narrow-mouthed water storage vessels and in situ chlorination in a Bolivian community: a simple method to improve drinking water quality. *American Journal of Tropical Medicine and Hygiene* 54: 511–516.

Rowe, A., Angulo, F., Roberts, L. and Tauxe, R. (1998). Chlorinating well water with liquid blech was not an effective water disinfection strategy in Guinea-Bissau. *International Journal of Environmental Health Research* 8: 339–340.

Srivastava, A., Nagpal, B. N., Saxena, R. and Sharma, V. P. (1999). Geographic information system as a tool to study malaria receptivity in Nadiad Taluka, Kheda district, Gujarat, India. *Southeast Asian J Trop Med Public Health* 30: 650–656.

Teixeira, J. E. (1968). Geologia da Guiné Portuguesa. *Junta de Investigação do Ultramar* 1: 53–104.

UN (2006). *The Millennium Development Goals Report.* New York: United Nations Department of Economic and Social Affairs.

UNDP (2006). *Human development report.* New York: Palgrave Macmillan.

UNICEF (2006). *Progress for children.* A report card on water and sanitation No 5. New York: United Nations Children's Fund.

WHO (2005). *Cleaning and disinfecting wells in emergencies.* Tecnhical notes for emergencies N.° 1. Geneva: World Health Organization.

WHO (2006). *Guidelines for Drinking-water Quality, First Addedum to 3rd edition*, Volume 1. Geneva: WHO Press.

WHO (2007). *Combating waterborne disease at the household level / International Network to Promote Household Water Treatment and Safe Storage.* Geneva: World Health Organization.

WHO/UNICEF (2006). *Meeting the MDG drinking water and sanitation target: the urban and rural challenge of the decade.* Geneva: World Health Organization and United Nations Children's Fund.

Witt V. M. and Reiff F. M. (1993). *Estrategias Alternativas Para la Desinfeccion del Agua Aplicable a Zonas Urbanas Marginales y Rurales.* Washington, DC: Pan American Health Organization; May 1993, Serie ambiental 13.

World Bank (1997). *Guinea-Bissau water and energy project*, Report No. PIC1858, The World Bank, Washington, DC: World Bank.

Wright, J., Gundry, S. and Conroy, R. (2003). Household drinking water in developing countries: a systematic review of microbiological contamination between source and point-of-use. *Tropical Medicine & International Health* 9: 106–117.

Zar, J. H. (1998). *Biostatistical analysis*, 4th Edition. Upper Saddle River, NJ: Prentice-Hall Inc.

Godwin Uyi Ojo*

The politics of water management in rural communities of Nigeria: Deep or shallow commitment?

Abstract. This paper assesses the politics of water management in relation to the institutional context leading to water privatization and commoditization in Nigeria. The near collapse of public utilities in water supply has led to the deployment of integrated water resource management to meet water needs. Several efforts, including the privatization of water supply, have been made to provide water to the population especially in the rural areas where over two-thirds of the people live. In spite of these efforts, the results show that potable water supply remains a pipedream. The paper assesses the institutional framework and the context in which clean water provision has been pursued, and argues that the failure of public water supply has led to water privatization rather than direct efforts by governments to privatize the resource. The comparative study examines the external actors' community participatory water resource management initiatives, and the relations of power of the social actors in the development process, which have significantly compounded the politics of water supply. Drawing from semi-structured interviews, direct observation, and multiple case studies, the paper argues for an enhanced integrated approach for catchment areas rather than the emerging proliferation of water stations by private and public enterprises.

Keywords. Actors, commoditization, participatory development, water supply, Nigeria

* Sessional Lecturer, Department of Geography, School of Social Science and Public Policy, King's College London, Strand, WC2R 2LS. Email: Godwin.ojo@kcl.ac.uk.

Introduction

Notwithstanding its shortcomings, there is a growing consensus among scholars and practitioners that integrated water resource management (IWRM) presents a holistic approach towards clean water accessibility and sustainable consumption (Molle, 2009; Akpabio et al.; 2007; GWP, 2008; Cleaver, 1999; Salau, 1986; Adams, 1985). The authors identified the chief obstacle as the low-level institutional operation embracing wide ranging issues of governance and decision-making, levels of participation of the actors as well as financial commitment. The basis of this assertion also rests on the consideration of integrated water management model that lays claims to the maximization of economic and social benefits in a manner that does not mortgage the lives of future generations.

In diverse guises, public participation and participatory approaches aimed at resolving the differing interests and world-views of the multi-stakeholders in water management have been married to integrated water scheme practices. Some scholars promoting this approach emphasize the need for cooperation across multi-scales bringing together government officials, policy makers, implementers and the active participation of user groups for effective water management (Molle, 2009; Akpabio, 2008a; Adams, 1991 and 1985). A recent effort aptly summarized this interface in an alphabetical mimetic device, *ABCDE* referring to: assessment of water availability; bargaining and choice making; codification of laws of cooperation; delegation of authority for project implementation; and engineering in relation to infrastructure provision (Jagannathan et al., 2009). Such an approach involving complex and integrated processes of cooperation helps to reinforce the notion that politics across spatial and temporal scales is central in water management.

The wide range in actors, developmental variables and indeed levels of development, country to country and community to community led Biswas (2004) to question the integrated water resource management approach based on place differentiation, and its dogmatic, one-size fit all façade. For centuries, freshwater supply in African rural communities, particularly in sub Saharan Africa, has depended mainly on traditional sources of water supply. In general, geographical studies of settlement formation showed that they were mostly located in proximity to natural watercourses such as rivers, springs, lakes or wells. However, as the population and communities have grown the challenge of water supply in both small and large communities has continued to give rise to diverse approaches to groundwater schemes (Henderson-Sellers, 1979). Such schemes for balancing supply and demand usually in economic terms have become a feature of water supply from aquifers, subterranean reservoirs, rivers, and dug-out wells. In contrast, amongst development practitioners, who are external to the communities providing water aid, piped water supply is often assumed to be improved whilst other forms of technology are classified as unimproved (Page, 2009). The author noted further that in the tropics, properly maintained hand-dug wells or springs can be safer than poorly maintained boreholes or piped reticulation systems (see also Black, 1990).

Notably, the narratives that have directed water management are socially constructed via a process that is laden with politics. This has to do with the way in which resource management issues are framed so that the problems and remedies that gain visibility are often products of the relations of power (Castree, 2001; Bryant et al., 1997; Escobar, 1995). Similarly, the rhetoric of integrated river basins is driven by "a political and ideological construct", with discursive representations and justifications related to governance, ecological, and economic considerations (Molle, 2009: 492). Thus, while such categorizations of traditional means of water management tended to erode the relevance of the long history of traditional water management, in fact, communal institutions and rules of behaviour around watercourses and springs are notable (Page, 2009; Adams, 1992; Kukah, 2003; Ostrom, 1990). Again, such mainstream narratives tended to promote exogenous and sophisticated technologies that are even less environmentally friendly.

Inevitably, such barrages of narratives are inadvertently or purposely at the forefront for championing the process of water privatization and commoditization. This shift in water management has a corresponding trend in the politics surrounding water narratives and, thus, influences the aggressive transformation of water supply systems from dependence on context based systems to more sophisticated yet unproven universal ones promoting large dams and reservoirs.

That notwithstanding, the importance of integrated water management its nature as a means to an end as the quest for rational water management may contribute to the alleviation of poverty, maintenance of ecosystems and conservation of natural resources (Akpabio et al., 2007; Mitchell, 1989; Salau, 1986). Yet, this chapter looks beyond IWRM to highlight the challenges posed by participatory water resource management, the contextual factors, and the intended and unintended consequences associated with this choice. The chapter is divided into three sections. Section 1 takes a look at the institutional framework for water supply in Nigeria and its attendant policy inconsistencies which portend flashpoints of conflict. Section 2 focuses on cases of participatory

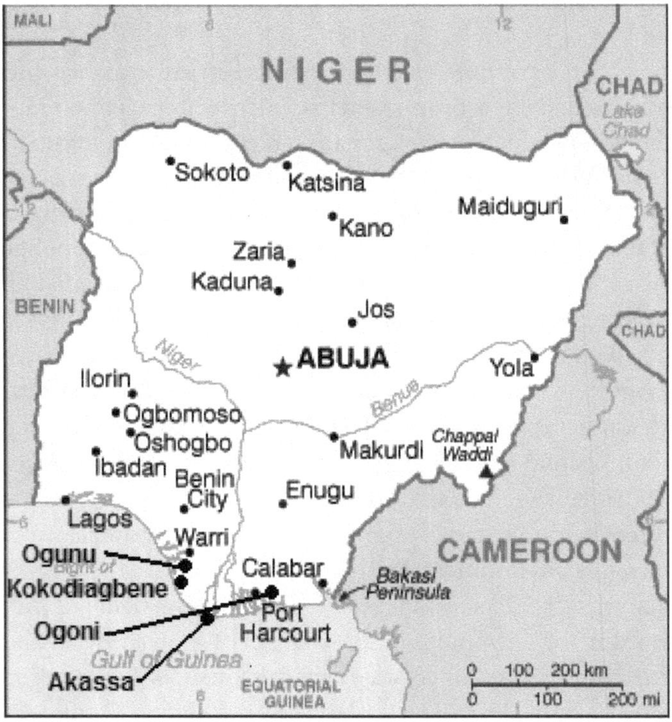

Figure 1: Map of Nigeria Showing Ogoni and Akassa Communities.
Source: The World Fact Book (2006).

approaches to community water resource management initiatives by a Transnational company and an NGO to illuminate contextual problems in water management initiatives by external actors. Section 3 deals with the emerging issues in water management in relation to local perceptions of roles and responsibilities and the implications for fresh water delivery. The concluding section draws the parts together and questions the onus of thrusting water development and delivery at the domain of the individual and underpinning the intended and unintended consequences of the external actors' participatory water delivery efforts.

Water rights and institutional framework

Nigeria's 140 million people make the country the most populous in Africa. It lies in the tropics located in the Gulf of Guinea, West Africa. It

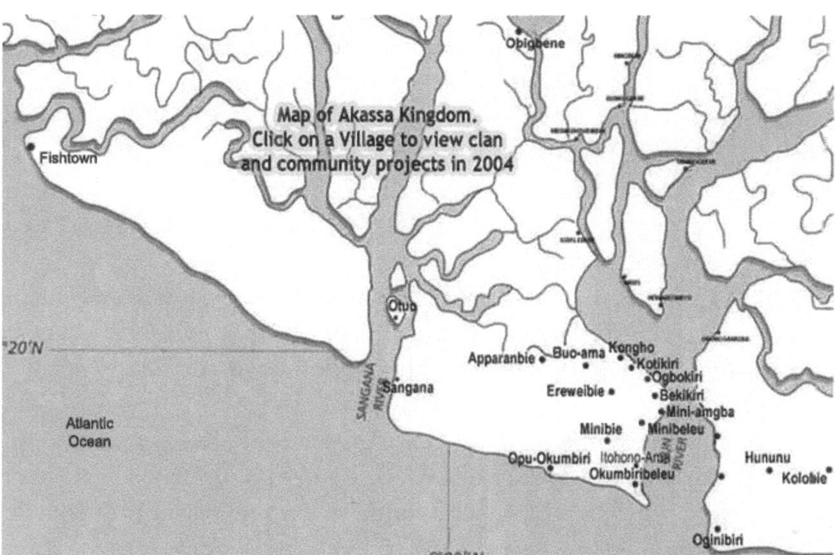

Figure 2: Akassa Community: Source: PNIN (2007).

is between the latitudes and longitudes of 10 degrees north and 8 degrees east respectively, with a total landmass of about 923.768 km^2 (see Figure 1). The main geographic feature is the River Niger and River Benue which drain 60 percent of the land area and form a maze of delta in the south. The gradient is gentle with a gradual loss of altitude from about 500m in the north to the coast. It is highly seasonal with the wet season stretching from June/July to August/September and the rest of the year being dry. The average annual rainfall (500mm) is derived from a varied regional spread of over 4,000 mm in the southeast to below 250 mm in the extreme northeast.

Securing water rights for the teeming population has been an uphill task. It is directly related to the institutional framework in place to guarantee the enjoyment of the resource. Efforts toward a national integration of water policy rarely take contextual factors into account when planning projects and development. In Nigeria, such local differences across regions and communities especially regarding local values and water availability, have been identified by Biswas (2004) as having adversely affected the impact of IWRM schemes. For example, experiences from differing regional River Basins show that a weak institutional framework threatens the efficacy of IWRM particularly due to legal, administrative and financial constraints; the result is the inability to provide the conditions for effective water management (Molle, 2009; Akpabio et al., 2007; Mitchell, 1994; Adams, 1991; Salau, 1986; Okafor, 1985). The multi-interest, multi-agency and multi-stakeholder dimensions to integrated water resource development posed institutional problems that are often linked with complicated systems (Akpabio, 2008a).

At the policy level, the initial efforts at integrated water management were set by the Federal Government of Nigeria (FGN) when it promulgated Degree 1976 establishing two river basins (subsequently amended in 1987), which resulted in the establishment of nine additional River Basin Development Areas (RBDAs). In 1985, the Directorate of Foods, Roads and Rural Infrastructure (DFRRI) was established by military decree No. 4 of 1985 with special emphasis on the delivery of potable water in communities where these services were lacking (Olayiwola et al., 2005; Black 1990). However, political patronage, limited funding and poor execution robbed the population of positive gains in this concern (Olayiwola et al., 2005; Adams, 1991).

Meanwhile, the Land Use Act of 1978 contradicted the final authority of the FGN and gave state governments, local authorities, and even individuals the authority to exploit water (Akpabio, 2008b; Ojo et al., 2005). Similarly, Decree 101 of 1993 vested all rights of ownership and control of water in the FGN. Drawing from the prevailing land tenure rights, it also specified, though equivocally, that individuals are allowed unlimited access and without charge to groundwater exploitation for domestic use, watering livestock, or for fishing from any watercourse to which the public has free access.

There has been some degree of inconsistency at the policy level which has had a direct impact on practice. Firstly, while the regulatory provisions recognised the universal right to water, they did not, however, guarantee its functional provision. Secondly, the policy provisions suggested that water rights were based on land rights insofar as the right to land occupancy protected such rights. Thirdly, the growth in the scope and number of institutional frameworks complicated the process of cooperation among the diverse agencies operating at the federal, state and local government levels. In the main, since the land use act vested authority of lands on the 3-tiers of government, this provision heightens the real and potential conflicts between governmental agencies. It also instigates conflict between individuals, many of whom asserted ownership without deeds but according to traditional land rights (see also Page, 2009). Other inconsistencies from these provisions and the land use act have been highlighted including a lack of measures for defining "accountable access" by individuals, corporations, and governments (Akpabio, 2008a: 274). The situation of conflict, arsing from policy flux, is the direct result of the ambivalence of government over its role in water supply for domestic consumption. We will revisit this topic later in this chapter.

Policy contexts in practice have favoured large-scale projects over small ones even in the lack of evidence. In a case study involving UNICEF collaboration with the Nigerian government to deliver a low-cost technological approach to water supply, it was difficult to convince decision-makers to accept that hand pump schemes rather than boreholes were the right approach for relatively small and often remote communities

(Black, 1990). The author noted that water supply systems were perceived erroneously mainly in terms of "reservoirs, piped schemes and treatment plants; and sanitation conjured up mains sewerage, septic tanks and flush toilets" (Black, 1990: 91). Where precipitation and runoff is relatively stable multi-purpose tube wells with functioning hand pumps have been successfully tested in pilot projects in some rural communities of Nigeria, India, Bangladesh and Pakistan (Meinzen-Dick, 1996; Black, 1990). That notwithstanding, the consideration of adaptable technologies and local imperatives in water management has yet to gain wide acceptance within the government which prefers large-scale projects.

In contrast, the basin approach is noted to promote large-scale dams, irrigation and hydroelectric power generation as economic benefits but with high hidden costs (Molle, 2009; Adams, 1991). Initially guided by a welfarist orientation of non-profit, the government took responsibility for water supply through the river basin approach to end users at a highly subsidized rate (Salau, 1986; 1990). Up until to the 1980s, this approach led the government to build numerous large multi-purpose dams for improved irrigation, fisheries, hydro-power, and domestic supply. At that time, the central government's focus was mainly on urban centres. However, in general terms, piped water gradually became the hallmark of development even in rural areas. Government water agencies provided piped water to public taps via networks of pipelines from the water reservoirs to strategic locations and homes. However, the lack of dam maintenance led to the serious deterioration of the water provision system.

Some exogenous factors influenced the policy direction of Nigerian water services. The global free-fall in oil prices led to Nigeria's economic crisis in the mid-1980s and high external debt led to imposed macro-economic reforms. The country faced the International Monetary Fund's structural adjustment recipe that rationalised the public sector and cut public spending on social services. The cash-strapped economy witnessed apparent stagnation and even erosion of water provision facilities. Unlike the experience of some African countries (e.g. Cameroon) where public taps became privatized through commercialization (Page, 2009), in Nigeria, they simply vanished from the landscape, systematically removed by government utility board officials. This change in water management towards commercialisation saw an upsurge in the demand and installation of piped water to homes. The importance of this change indicates that government seemed mindful of the people's opposition to water privatization and commercialization. Nevertheless, the government had yielded to the strong wave of marketization of water and globalization that were externally driven on the wings of the neoliberal economic paradigm promoting them.

Faced with reduced funding, the water facilities with low level maintenance and know-how simply collapsed and the poor service aggravated the acute water problems witnessed in the upland areas since the 1980s. Distrust of the government's ability to provide adequate services heightened water pessimism. In government circles, the situation brought two opposing paradigms into collision: those seeking private sector participation and those advocating community participation. On the one hand, the reality of poor service places water development and delivery in the hands of individuals and communities. Government had no viable alternative other than to statutorily recognise individual water rights. On the other hand, this situation also created the impetus for the central government to limit itself to regulatory and monitoring roles through a policy and legal framework that allowed for deregulation, commoditization and competition. In practice, the granting of water licenses to commercial interests obfuscates government's responsibilities in the development and delivery of water to citizens.

The inconsistencies in uncoordinated water governance frameworks both in theory and practice resulted in institutional conflicts that placed water ownership and control in conflict. The institutional flux also signalled a shift in water policy from a centralised mode to a quasi-decentralized one involving third parties: individuals, communities, and commercial interests. This orchestrated a move from perceiving water as a public good towards water privatization with commercialization as its consequence. In particular, the shift also re-enforced community participatory development processes in water resource management led by external actors. External actors seemed to have strategically deployed water provision in the communities and this impinges on the level of their commitment.

External actors' participatory water resource management

Participatory development in community resource management gained wider popularity in the 1980s. Since it has some elements of communal practices, it is not entirely new in Nigeria. Earlier accounts from the 1940s describe participatory strategies deployed by the British Colonialists in their efforts to provide roads, water, and other basic social services (Chadwick, 1948). In the 1980s and 1990s large scale water projects and dams construction laid claims to participatory approaches. However, they were carried out perfunctorily, and the projects were subjected to high profile political patronage and the misappropriation of government revenue (Adams, 1991). However, the vacuum created by the government's inability to meet water demands for domestic and industrial uses paved the way for third party (such as transnational corporations and NGOs) involvement in community water projects.

These external actors laid claims to participatory development in community resource management that would allow greater community participation in project planning and management for optimal benefits. Indeed, local participation in project planning is gaining wider acceptance and replacing the top-down orthodoxy prevalent before the 1980s. It is now an indispensable formal component of development policy, planning and management. The Brundtland report and a number of other high profile international interventions indeed specify the right of local people to participate in sustainable development with the notions of citizenry and social justice embedded in contemporary democracy (WCED, 1987; Redclift, 1992).

Some scholars express the hope that participatory development will deliver on the goal of sustainable community resource management and local empowerment if development practices are based on local knowledge generated by intended beneficiaries (McGee et al., 2000; Chambers, 1983). According to Chambers (1997: 117), PRA allows local people to receive the "stick of authority", whilst becoming the analysts and planners themselves.

Other scholars showed less optimism for participatory development as an answer to the overarching problem of community resource management. Although PRA has gained institutional currency in participatory development practices sponsored by governments, donor agencies, NGOs, transnational corporations, the World Bank and other development agencies, the manner in which the process may be manipulated and subject to abuse has been documented (Hickey et al., 2005; Cooke et al., 2001; Blackburn et al., 1998). It is noted that there is a wide gap between the "grand rhetoric and the day-to-day practice of participation" (Blackburn et al., 1998: 6) and that the approach can be superficial, manipulative, and tyrannical to the extent even of disempowering those it was meant to empower, and thus contradicting its rhetoric of empowerment; in other words, it can "facilitate the unjust and illegitimate use of power" (Cooke et al., 2001: 3).

This concern for power and agency visible in critical development studies highlights the problems and constraints of participation as actors attempt to influence and control development planning processes that mediate human-environmental interactions (Robbins, 2004; Ojo et al., 1998). They do so by anticipating the agency and motivations of other actors even as they re-adjust behaviour and strategize to negotiate meanings in contexts "wherein actors create space for their own 'projects'" (Long et al., 1992: 10). Power relations thus invariably impinge on participatory outcomes.

The external actors' initiatives for community water supply have been pursued within the remit of participatory development. This research used semi-structured interviews and direct observation; its interview questions related to local perceptions regarding water management, actors' roles and responsibilities and the notions of water as a social service versus commoditization. It found that the level in which participatory development in water provision was deployed in the rural communities was indeed a product of power relations.

Case study I: Pronatura

Pronatura International, Nigeria (PNIN) is an NGO involved in participatory community resource management and was founded in 1992 in response to the influence of the principles of the Rio summit held in Brazil that same year. It operates extensively in the Niger Delta communities and combines rural development with turtle conservation in the Akassa communities, located on low land tropical rainforest on shore islands (see Figure 1). As Figure 1 shows, the area is contiguous to the Ogoni community that is another focus of this study. Akassa was the centre for palm oil trade for the Royal Niger Company built by the English adventurer and entrepreneur George Taubman Goldie in the 18th century (Don Pedro, 2006). The region is also noted for corn and rice farming in the wetlands. The local population is about 40,000 and is made up of several small clans and settlements (some residents claim there are over 80,000 people) (PNIN, 2007). The people are predominantly fisher folk, exploiting the diverse sub-ecozones to catch crustaceans such as periwinkles, crabs and oyster from the mangroves as well as some deep-sea fishing (Ashton-Jones et al., 1998; Alagoa et al., 1989). There are also endemic species of turtles (e.g. leather back and hawksbill), which are highly endangered but also provide protein for the local communities. The presence of such species was a key reason for PNIN's decision to locate projects in the area.

To a large extent, PNIN's participatory efforts in water provision have been influenced by community experience rooted in political and economic considerations. In general, participatory approaches to community resource management in the PNIN-led Akassa community projects has been largely successful in facilitating bottom-up approaches to community development planning to the extent that residents have a say in the choice of community development project planning and management. The establishment of the Akassa Development Foundation in 1997 set the stage for community participation in project planning and management and resulted in the construction of basic infrastructure such as roads, bridges, jetties, and cottage hospitals.

Indeed, the PNIN led participatory community development through its water and sanitation programme which was instrumental in re-establishing confidence in the local practices governing water sources. This encouraged the protection of the community wells by activities such as weeding around them and stimulated a campaign against unsanitary conditions whilst drawing attention to the health implications (PNIN/ADF, 1998).

Benefits notwithstanding, the construction of social realities regarding water within the framework of commodification seemed externally imposed and has resulted in non-cooperation and even rejection. Economically, commoditization and commercialization of water does not guarantee equitable access to water resources since some residents are poor and members of vulnerable groups who will never be able to afford the tariffs (Gelachi et al., 2009). Geographically, the Akassa communities are in the low-lying Niger Delta tropical monsoon zone. Thus, water precipitation reaches 4000 millimetres per year (Ashton et al., 1998). Still, its floodplains are drained by natural seasonal floods, networks of creeks, lagoons, and perennial streams. The location as a coastal settlement also meant that the water table is closer to the surface averaging approximately 20–30 metres (Black, 1990). It is common knowledge that the water problem in Akassa was not that of water availability but that of water quality.

During the participatory surveys by the communities organised by PNIN, poverty alleviation initiatives were restricted to livelihood and infrastructural projects and without focusing on water needs. Such local pragmatism meant that some pressing matters were left off the agenda. Thus, the community acquiesced in the denial of clean water as a priority. During the participatory appraisal, the water issue had indeed been a major problem identified in most communities but it kept coming on and off the list. Indeed, to the Kongho community residents, access to clean water is a problem in Akassa and they found it quite unfortunate that water needs were not currently being addressed by PNIN. The need for potable water, then, was important and even today remains a top priority that the community would like to pursue.

Yet, this external neglect of the water issue was not new but has greater political and economic significance. Between the early 1980s and 2005, the Kongho community had at least three different water projects commissioned by the state government and other development agencies, but none of these projects is yet functional. Generally, in Kongho, water projects are waste pipes through which money for community projects are either embezzled or misappropriated. Here, the high costs associated with water projects and the failure of such projects in the past combined so as to influence local decision-makers to exclude water as a priority. In the end, through the participatory development process, water was denied priority.

The need for clean water supply in Akassa and throughout the Niger Delta is not only high but is compounded by pollution and poor sanitary conditions. Although water abounds in the deltas, the risk of pollution from oil spills and industry makes it unsafe for drinking (Ashton et al., 1998). Under these circumstances, exploiting ground water sources may become increasingly difficult and uneconomic if alternative suitable surface water sources are available (Molle, 2009; Akpabio, 2008a). This may imply that the harvesting of rainwater through intermediate technologies rather than sophisticated systems will most likely be the solution for sustainable fresh water supply in these delta communities. It is argued that countries with multiple sources of water supply such as renewable surface water (e.g. large rivers), and groundwater sources of supply do relatively better in water resource management than countries who depend only on groundwater (Jagannathan et al., 2009). However, such a merry-go-round posture may not necessarily be the result of effective institutional mechanisms but rather the weak control mechanism that renders the over-exploitation of water resources difficult to control. Sustainable use of water resources requires careful weighing of supply options and choice-making that best suits the people and context.

Ten years down the line however, the PNIN and the ADF in conjunction with the European Union (EU) eventually prioritized water. Consequently a borehole was commissioned and built in Kongho in 2006. The project was jointly financed in a counterpart funding mechanism that saw the locals contributing about 20 per cent of the total costs. Local input was not in monetary terms but in the provision of designated community land, labour, sand and other construction materials. In spite of local institutional mechanisms for the water project management via the relatively strong ADF, the project functioned for less than one year from its commission. Even when it was functional, water quality was described by residents as odious and chalky perhaps due to poor engineering.

From these examples, it is noteworthy that the local institution, ADF, failed to contribute some of its cooperative funding to the water scheme. This suggests that the Kongho community residents are less willing to commit to commodity water projects and are not prepared to shoulder the burden of operational and maintenance costs that are located in water tariffs and cost recovery systems.

In spite of PNIN's facilitating community empowerment and the building of local institutions to manage community development, participatory development led by the NGO has not resulted in the efficient delivery of clean water. Indeed, participatory development became a mere buzzword (Cornwall et al.; Brock, 2005) mutually deployed to deny the reality of clean water needs in Kongho. This has had serious implications for clean water provision for rural dwellers.

Case study II: SPDC

Similarly, the Shell Petroleum Development Company's (SPDC) initiatives in community water management also lay claims to participatory process. The company even acknowledged that the shift from the Community Assistance to Community Development approach meant the need for communities "to take the lead role" in their own development (SPDC, 2004: 15; see also SPDC, 2007). By being thus an integral part of the intervention strategies, a participatory approach and 'participation' itself became a strategic tool for wooing the community to join in Shell's community development plans. Shell's initiatives in the areas of its operations also included "drilling deeper water boreholes" and "supported the rehabilitation of existing water schemes to improve their delivery" (SPDC, 2004: 21). The Bodo water project was established for the provision of water to the intended beneficiaries in the local community.

In contrast to PNIN, Shell is a commercial oil mogul and is the operator of the NNPC/Shell/TotalFinaElf/Agip Joint Venture with 30 per cent controlling shares. Clearly, Shell is a major player in Nigeria's economy producing on average one million barrels of oil per day and contributing about 43 per cent of total oil revenue received by the government (Shell, 2004a: 16). It is also by far the largest polluter of the environment and degrader of traditional sources of livelihoods in communities where it operates. Environmental problems here result from oil spills, pipeline blowouts, gas flares and deforestation (Doyle, 2002; Frynas, 2000; World Bank, 1995).

The Ogoni population is about 500,000 and these people live on approximately 1,000 square kilometres of land east of Port Harcourt (see Figure 1). Shell took 634 million barrels of oil from Ogoniland valued at about US$5.2 billion between 1958 and 1993 (Detheridge et al., 1998; Frynas, 1998). Yet, as Frynas (1998) points out, within this period only an inconsequential amount of Shell's profits from oil revenues from Ogoniland – about 0.000007 percent – was invested in local communities (Frynas, 1998: 463). The Ogonis themselves still lack electricity, pipe borne water, and basic social amenities.

Shell's deployment of participatory approaches in managing community pressure took a different turn when threats to its oil operations were at stake. By the mid 1990s, the company was under severe pressure to justify its continued presence in Ogoni given its considerably adverse impact on both the environment and the population. Scholars mentioned above document that a series of community protests against Shell culminated in the arrest and execution of the Ogoni activist and environmentalist, Ken Saro Wiwa, and eight others. Community residents claimed that, as a response designed to dampen community agitation and public outcry at the local and international levels, Shell introduced a few projects in quick succession as its contribution to community development in Ogoni. Significantly, though, community members were not aware of any participatory study for development planning in Ogoniland related to water and labelled the Bodo water project as part of the firm's peace gesture after the death of Ken Saro Wiwa in 1995.

The water facility project in Bodo which began in 2005 and was completed in 2006 was, above all, an exercise in managing company-community relations so as to reduce local tensions. Although the facility was an abandoned project and had been the responsibility of a defunct government agency called the Oil Minerals Producing Area Development Commission, Shell's involvement by some community members was a welcomed development. From a personal observation, the facility consisted of two large submersible pumping machines and a power plant housed in a bungalow. Another newly renovated twin flats bungalow provided the accommodation for the two staff members operating the facility. The water tank had the capacity to store about 100 cubic metres (one million litres) suspended on giant steel pipes about 30 metres high. From there, water was distributed to designated points in the community. The project is relatively big as far as a small community water project goes. Community members informed me that it cost several million Nairas, although there are no published costs. In the few weeks the water facility operated, people were happy. The demand was so high that water was pumped twice a day.

However, Shell's intervention raised questions about its commitment to meaningful community participation in project planning. The SPDC contractor exclusively appointed from nearby city of Port Harcourt, held consultations with the community chiefs and then a few weeks later, work began on the water project. Indeed, community involvement in the management process and in specifying roles and responsibilities also received zero attention.[1] The implication of this community exclusion from participation meant that the management of the project was left entirely to the contractor and without local people or committees to join in this process. The contractor, thus, also had the task of buying diesel for running the generator, employing staff to pump water from the borehole, and plant maintenance. Hence, local involvement consisted only in the provision of paid labour: no more than hiring the local youths to dig and lay the water pipes.

Not surprisingly, the plant facility is facing problems because the project has not been properly handed over to any local institution. When the contractor left, the project simply crumbled. Water leaked from the pipes either due to too much pressure or as a result of undersize pipes having been used. The power generating plant was also vandalised and was not repaired by Shell. Hence residents were back to where they started from: suffering from a lack of water. The local spring is not only several kilometres away but traditions and taboos have excluded men from washing in it.

The community hands-off approach in watching water projects overgrown by weeds relates to the mind-set involved in viewing Shell's projects as inroads to the exploitation of their crude reserves. Amongst these actors, social interactions involved a counter-balancing act of power. The way water issues were framed was couched in local politics, a fact

1 In a move towards local involvement, Shell facilitated the establishment of Project Management Committees (PMCs) in 2003 as local managers of its Bodo Agriculture Farm established by the company in 1977. However, the PMCs mandate excluded water management.

which has wider implications for the actors beyond the local community. Participatory development in water projects became a political tool by which Shell sought prior consent for the firms' oil activities as well as a means by which community residents expressed dissent to oil exploitation. As in the Akassa case, the non-action posture reveals the Bodo community resistance to water projects that sought to shift the burden of water delivery and costs of maintenance from the doorstep of government or external actors to individuals. Here again, water is seen as a social service that should be rendered freely by governments or its partners (especially Shell) to the citizens. The dire need for clean water has opened up opportunities for local elites who have sunk private boreholes, for a gradually booming business of the sale of water to the public.

Patterns of emerging issues in water management

The implications of the ambiguous legal framework that recognises universal individual water rights yet without guaranteeing them has given rise to issues in water management. Firstly, there is confusion over roles and responsibilities in water delivery. Table 1 shows communities' perception of stakeholders' roles and responsibilities in water management. At least at the institutional level, the role of government in water delivery beyond the regulatory framework is unclear. On the part of the communities, there are also conflicting signals over water ownership rights, roles and responsibilities. The numbers and percentages overlap. While only five per cent of interviewees deferred water ownership rights to the government by virtue of its role as policy-making at the macro-level, the majority (95 per cent) asserted community ownership to water resources. Yet, 95 per cent of community members declined the responsibility for water development and delivery because most residents thought water development and delivery was the primary responsibility of government to her citizens.

However, 80 per cent of the community interviewees were willing to take full responsibility only if statutory allocations directly supported local initiatives. Meanwhile, there seemed to be a consensus that the

Actors	Local perceptions on water ownership	Rights and instruments	Roles and responsibilities
The 3-tiers of government (Federal, state and local government authorities)	Own water resources and land (5 %)	Land Use Act 1978	Delivery of free water supply for domestic consumption (95 %)
Communities	Community owns water resources (95 %)	Community customary rights (no deeds)	Community to provide water through statutory provisions from government or donor agencies (80%)Community unwilling to shoulder water costs (95 %)
Individuals	No individual or companies should lay claim to natural watercourses for sale (Consensus)	Free gift of God	Community members are willing to donate labour and materials but will not make cash ommitments (Consensus)

Table 1: Showing local perceptions of water ownership, roles and responsibilities.

individual right to water remains subordinated to community rights and control. In the rural communities water is seen as more than a public good, rather as the source of life and a free gift of God. Hence, water should never be appropriated for commercialization by the government, companies or individuals. Such aversion to water commercialization and trade stemmed from the conviction that the buying of water throughout life for cooking, drinking and washing amounted to being in perpetual slavery at the mercy of water vendors. Yet, a significant number, five per cent, said that they would resort to buying water if it was proven to be cheaper and delivered higher quality than other available sources. Indeed, dependence on fresh water supply from springs, rivers, wells and runoff in the community has been complemented by imports of bottled water from bigger cities by commercial interests. While the community is adverse to water commoditization, yet marketization is increasing as is patronage of water vendors.

Further, the lack of an institutional framework to guarantee clean water to the public provided the space for the involvement of external actors. In theory, the policy framework tended to promote water as a free public good by ensuring that everyone even without a licence could exploit freshwater aquifers. Contrarily, in practice, the implementation of this policy contradicts rather than complements in the contentious issues of water ownership, control and management. Indeed, the changing land tenure concepts are complex and combining government policy with customary practices can, in the hands of elites, serve as a means for manipulating the systems to gain control of land, water and other vital resources important to the collective (Palmer-Jones, 1993; Tiffen, 1985). Conflicting water policy is a hallmark of sub-Saharan Africa.

Secondly, evidence suggests a gap between community organisations and water delivery in the rural communities. The dominant role in community resource management (e.g. water and sanitation schemes) is increasingly growing in importance (Opare, 2007; Hulme et al., 1997; Black, 1990). Yet, community based institutions such as the PNIN's ADF or the Shell's PMCs, though working in the right direction, are not adequately supported to conduct community development on their own terms. In relation to water, there is little commitment on the part of both the external actors and the communities. Local institutional structures taking responsibility for water development and delivery have been found to be a key element as was the case in the UNICEF and Nigeria collaboration in the 1980s (Black, 1990). Participatory approaches in water management are gaining ground and women's traditional role as domestic managers has provided the obvious rationale for their involvement in water operation, maintenance and in decision-making (Page, 2009; Wijk-Sijbesma, 1985).

Thirdly, water privatization has culminated in commercialisation which has resulted in unintended disparity in assess and use between the rich and the poor as well as providing considerable benefits in terms of returns on investment from the informal markets. The new water merchants are not only national or transnational brand names but an increasing number of local elites able to afford boreholes and sell water in packages, tanks, and sachets. The implication of this type of deregulation by default initially in the urban areas implies that, across the cities and villages, anyone able to afford boreholes can install such water stations in their homes.

Indeed, the construction of water stations can be burdensome due to the high costs of the purchase of the machinery, installation and maintenance. Water stations consisted of boreholes drilled into ground water aquifers. Unlike the low-lying Niger Delta, the depth reaches from 80–100 metres especially towards the semi-arid zone in northern Nigeria. A submersible water pump and a medium private plant running on fossil fuel are required to pump water to storage water tanks extended upwards to 20 to 40 metres high. From this storage tanks water is distributed to households or strategic locations in the community. It is not environmentally friendly, and can be a source of noise to neighbours. In a nation where the national electricity grid is epileptic and inconsistent the costs of such water stations and there maintenance represent a huge financial burden, not least of all for the poor.

In urban areas, this approach has become the prevalent source of water. Indeed, almost every household that can afford to, now sinks boreholes in their compounds, and it is common to see a dozen boreholes in one locale. The proliferation of water stations is a serious cause for concern for groundwater management.

This pattern of deregulation has also effectively commercialised water in ways that some community residents adverse to privatization are forced to either sell or buy water from these private water stations. Further, local appetite for bottled water has increased and has become a sort of status symbol. It is now a common feature for guests during parties such as anniversaries, weddings, and festivities to expect to be served customized bottled water as entertainment and souvenir. Such acceptance is ironical. It is ironic that while people are adverse to water commercialization or even shifting the burden of maintenance of water facility, yet there are rapidly growing informal markets and trade in water in the rural communities as well as in the urban areas. Inadvertently, though, communities that are strongly opposed to water privatization and commoditization became the very instruments and inroads for promoting water commercialization.

Conclusion

This chapter has assessed the external actors' initiatives, the context of their involvement, and the extent to which participatory process significantly influenced the direction of water projects in the Akassa and Ogoni communities. From the communities, gauging the water management scale amounted to an exercise in pessimism. Indeed, many scholars who express pessimism argued that the exiting institutional framework for integrated water management constitutes a major barrier to effective implementation (Akpabio et al., 2007; Olayiwola et al., 2005; Cleaver, 1999; Adams, 1991). Yet, identifying how to evolve a coordinated approach to water management remains a daunting task.

The chapter suggests that the privatization of water has given rise to its commercialization. Yet, water privatization and commercialization was not due to a deliberate government policy but rather due to its collapse. Internally, sheer mismanagement and poor water delivery capabilities influenced this turn. The contending forces between communal and individual rights advocates and commercialization interests forced water privatization underground through the entrance of private entrepreneurs and individual initiatives to fill the vacuum.

Externally, a focus on integrated water management failed principally due to lack of consideration of local contexts. Given the peculiarities of the Niger Delta and the high rate of precipitation, the choices available at the local level have not been adequately explored to avoid what seems like wasting the rain (Adams, 1992). Indeed, low user participation in areas where there is less scarcity of water is the norm for water projects that are often based on financing and cost-recovery (Akpabio et al., 2007). Thus, exploring the abundant untapped runoff and rainfall harvesting with cheap and intermediate technologies seemed a better option than the sophisticated and expensive water stations. By genuine support the rural

Niger Delta can regenerate itself by focusing on how to convert water quantity to water quality. Relying on donations of labour and time, it can, even in this neoliberal climate, place management and delivery obligations on its citizens in communal arrangements that are often devoid of massive capita.

The institutional and legal framework needs to be streamlined in order to clarify its ambiguities. Although community members in Kongho and Bodo perceive water as a social service, and thus are adverse to its commercialisation, yet they have in some ways contributed to the water trade through rising demand, sales and consumption of bottled water. Reconciling these conflicting positions may mean greater focus on the direction of external facilitation of the local institutional empowerment. Taking into account local contexts in water project planning will also require strengthening the local communities to develop institutional strategies through local organising practices and modernisation possible within common property resource management regimes (see Ostrom, 1990).

Participatory process in integrated water management remains a viable option but not in the manner it has been strategically deployed to align external actors' organisational interests with community needs. This calls for greater involvement in project monitoring and feedback in participatory development. In the case of the NGOs, while the participatory process denied the reality of water as a priority for better conditions of living, that of Shell failed to take into account community participation in project planning and design as well as in management. Both external actors' initiatives failed to provide clean water because there was very little or no appreciation of integrated water management in the communities. The cases were inevitably that of shallow commitment that sinks deep in pessimism. The different cases suggest as well that actor-type is significant in the amount of rejection or acceptance of water projects in the communities but more study is required to document how this may affect local commitment in water delivery schemes.

The proliferation of water stations is ill-conceived in terms of wastage and inequitable access. It is unsustainable as it is the least affordable. Water privatization has culminated in the growing informal water markets and has played an important role in the expansion of unaccountable access to critical groundwater resources (Meinzen-Dick, 1996). Overexploitation of ground water resources is inevitable in the proliferation of water stations. Further study is needed to determine the impact of the subterranean water markets on the water table, and, on a broader scale, on the per capita consumption of water in the urban and rural communities to guide in water production and delivery. The unregulated private water stations which are fast becoming channels for the privatization and commoditization of water have a direct impact on the ground water aquifers. This can further compound the problem of sustainable livelihoods and water needs.

This chapter contends that only a holistic approach with attention to local contexts of water supply can help change the situation of hydro-pessimism to that of optimism. A focus on integrated water supply through a catchments approach should take the centre stage rather than the current situation of individuals becoming water vendors and suppliers to less privileged neighbours. Water supply should be regulated in a dis-centralised manner but through properly defined catchment management regimes to check water wastage and guarantee equal access.

Acknowledgements

This work benefited from comments made by Raymond Bryant and Ben Page on an earlier version. The views expressed are solely those of the author and do not represent the views of the reviewers or the institutions they represent.

References

Adams, W. M. (1985). River Basin Planning in Nigeria. *Applied Geography* 5: 297–308.

Adams, W. M. (1991). Large Scale Irrigation in Northern Nigeria: Performance and Ideology. *Transactions of the Institute of British Geographers* 16: 287–300.

Adams, W. M. (1992). *Wasting the Rain: Rivers, People, and Planning in Africa.* Minneapolis, London: University of Minnesota Press.

Akpabio, M. Emmanuel (2008a). Integrated water Resources Management in Cross River Basin, Nigeria: How can we reconcile institutional boundaries and interests? *International Journal River Basin Management* 6 (3): 267–276.

Akpabio, M. Emmanuel, Watson, Nigel M., Ite, Uwem E., and Ukpong, Imoh E. (2007). Integrated water Resource Management in the Cross River Basin, Nigeria. *International Journal of Water Resources Development* 23 (4): 691–708.

Akpabio, M. Emmanuel (2008b). Institutions and Water Management in Nigeria: Showcasing the Scales of Authorities and Conflicts, A paper presented at the 4th International Workshop on Hydro-hegemony (Power and Power Asymmetry in the Water World, London, 31st July.

Alagoa, E. J. and Tamuno, T. N. (1989). *Land and People of Nigeria: Rivers State.* Port Harcourt: Riverside Communications.

Biswas, K. Asit (2004). Integrated water Resources Management: A Reassessment. *Water International* 29 (2): 248–256.

Black, Maggie (1990). *From Handpumps to Health: The Evolution of Water and sanitation Programmes in Bangladesh, India, and Nigeria.* New York: United Nations Children's Fund.

Blackburn, J. and Holland J. (eds) (1998). *Who Changes? Institutionalizing Participation in Development.* London: IT Publications.

Bryant, R. L. and Bailey, S. (1997). *Third World Political Ecology.* London: Routledge.

Castree, N. and Braun, B. (eds.) (2001). *Social Nature: Theory, Practice, and Politics.* Oxford: Blackwell.

Chadwick, E. R. (1948). Mass Education in Udi Division. *African Affairs* 47: 31–41.

Chambers, R. (1983). *Rural Development: Putting the Last First.* London: Longman.

Chambers, R. (1997). *Whose Reality Counts? Putting the First Last.* London: Intermediate Technology Publications.

Cleaver, F. (1999). Paradoxes of Participation: Questioning Participatory Approaches to Development. *Journal of International Development* 11: 597–612.

Cooke, B. and Kothari U. (eds) (2001a). *Participation: The New Tyranny?* London: Zed Books.

Detheridge, A. and Pepple, N. (1998). A Response to Frynas. *Third World Quarterly* 19 (3): 479–486.

DonPedro, I. (2006). *Oil in the Water: Crude Power and Militancy in the Niger Delta.* Lagos: Foreward Communications Limited.

Doyle, J (2002). *Ridding the Dragon: The Royal Dutch Shell and the Fossil Fire.* Washington DC: A Publication of Environmental Health Foundation.

Environmental Rights Actions/Friends of the Earth Nigeria (2007). *Water Privatization in Nigeria*, <www.essentialaction.org/shell/era/era.html> (accessed May 25 2009).

Escobar A. (1995). *Encountering Development: the making and unmaking of the Third World.* Princeton: Princeton University Press.

Frynas, G. (1998). Political Instability and Business: Focus on Shell in Nigeria. *Third World Quarterly* 19 (3): 457–478.

Frynas, J. G. (2000). *Oil in Nigeria: Conflict and Litigation between Oil Companies and Village Communities.* Hamburg: LIT Verlag Munster.

Gelachi, Esther and Franceys, Richards (2008). Regulating Water Services for the Poor: The case of Amman. *Geoforum* 40 (3): 431–441.

Global Water Partnership (2008). *IWRM Toolbox Version* 2, Stockholm, <www.gwpforum.org> (accessed 15 July 2009).

Henderson-Sellers, Brian (1979). *Reservoirs.* London: Macmillan Press Limited.

Hickey, S. and Mohan G. (eds.) (2004a). *Participation, From Tyranny to Transformation?* London: Zed Books.

Hulme, D. and Edwards, M. (eds) (1997). *NGOs, State and Donors, New York and London: Macmillan Institute of Water and Environmental Management (1993), An Introduction to Drinking Water Quality.* London: IWEM Booklet 3.

Integrated Regional Information Network (2008), Nigeria: Bloody Week in the Niger Delta assets, <www.Irinnews.org/Report.aspx?ReportId =80502>.

Jagannathan, Vijay; Mohamed, M Ahmed, and Kremer, Alexander (eds) (2009). *Water in the Arab World: Management Perspectives and Innovations.* Washington DC: The World Bank.

Kukah, M. H. (2003). *Democracy and Civil Society in Nigeria.* Ibadan: Spectrum Books.

McGee, R and Norton, A (2000). *Participation in Poverty Reduction Strategies: A Synthesis of Experience with Participatory Approaches to Policy Design, Implementation and Monitoring.* IDS, Working Paper 109.

Meinzen-Dick, Ruth (1996), *Groundwater Markets in Pakistan: Participation and productivity.* Washington DC: International Food Policy Research Institute, Research Report 105.

Molle, Francois (2009). River-basin Planning and Management: The social life of a concept. *Geoforum* 40 (3): 484–494.

Ojo, G. U. and Ashton-Jones N. (1998). Community, Power and Conservation: The ERA Experience in the Niger Delta. In Daphe, T. (ed.), *Forests, Trees, and People*, p. 49–52. Newsletter No. 36/37, Uppsala.

Ojo, G. U. and Gaskiya, J. (eds) (2003). *Environmental Laws of Nigeria: A Critical Review.* Ibadan: KraftBooks.

Okafor, F. C. (1985). River Basin Management and Food Crisis in Nigeria. *Geoforum* 16 (4): 413–421.

Olayiwola, L. M. and Adeleye, O. A. (2005). Rural Infrastructural Development in Nigeria: Between 1960 and 1990 – Problems and Challenges. *Journal of Social Sciences* 11(2): 91–96.

Opare, Service (2007). Strengthening Community-based Organizations for the challenges of Rural Development. *Community Development Journal* 42 (2): 251–264.

Ostrom E. (1990). *Governing the Commons: The Evolution of Institutions for Collective Action.* Cambridge: Cambridge University Press.

Page, Ben (2009). The history of water politics and the politics of water history (Forthcoming).

Page, Ben (2003). Communities as the agents of commodization: The Kumbo Water Authority in Northwest Cameroon. *Geoforum* 34 (4): 483–498.

Palmer-Jones, Richard W. (1983). Mismanaging the Peasants: Some Origins of Low Productivity on Irrigation Schemes in the North of Nigeria. In A. T. Grove and W. M. Adams (eds), *Irrigation in Africa: Problems and Problem Solving*. Cambridge: African Studies Centre, University of Cambridge.

PNIN/ADF Angala News (1998). *Owei in Natural Resources Committee Meeting*. Kongho: Niger Delta Wetlands Centre/Pro-Natura International, No. 6.

Pronatura (2008), <www.pronatura-Nigeria.org>.

Pronatura International, Nigeria – PNIN (2007) *Organisational Profile*, Port Harcourt.

Redclift, M. (1992). Sustainable Development and Popular Participation: A Framework for Analysis. In Ghai J. and Vivian J. M. (eds), *Grassroots Environmental Action: Peoples Participation in Sustainable Development*, p. 23–49. London: Routledge.

Robbins P. (2004). *Political Ecology*. Malden: Blackwell.

Salau, A. T. (1986). River Basin Planning as a strategy for Rural Development in Nigeria. *Journal of Rural Studies* 32 (4): 321–325.

Shell Petroleum Development Company (2004). *2003 People and Environment Report*, Nigeria, 31 May.

Shell Petroleum Development Company (2007). *Shell Annual Report 2006: People and the Environment*, Nigeria, 8 May.

SPDC (2008), <www.shell.com>.

Tiffen, Mary (1985). *Land tenure Issues in Irrigation Planning Design and Management in Sub-Sahara Africa*. London, ODI Working paper No. 16.

Wijk-Sijbesma, Christine (1985). *Participation of Women in Water Supply and Sanitation, Roles and Realities*. The Hague: International Reference Centre for Community Water Supply and Sanitation, Technical Paper 22.

World Bank (1995). *Defining an Environmental Development Strategy for the Niger Delta Volume II*, Industry and Energy Operations Division, West Africa Department of the World Bank.

World Commission on Environment and Development (1987). *Our Common Future*, Oxford: Oxford University Press.

World Water Commission (2000). *World water Commission 2000: A secure Water World Vision for Water, Life and the Environment*. Marseilles: World Water Council.

PART IV
Hydropolitical risks and transboundary water politics

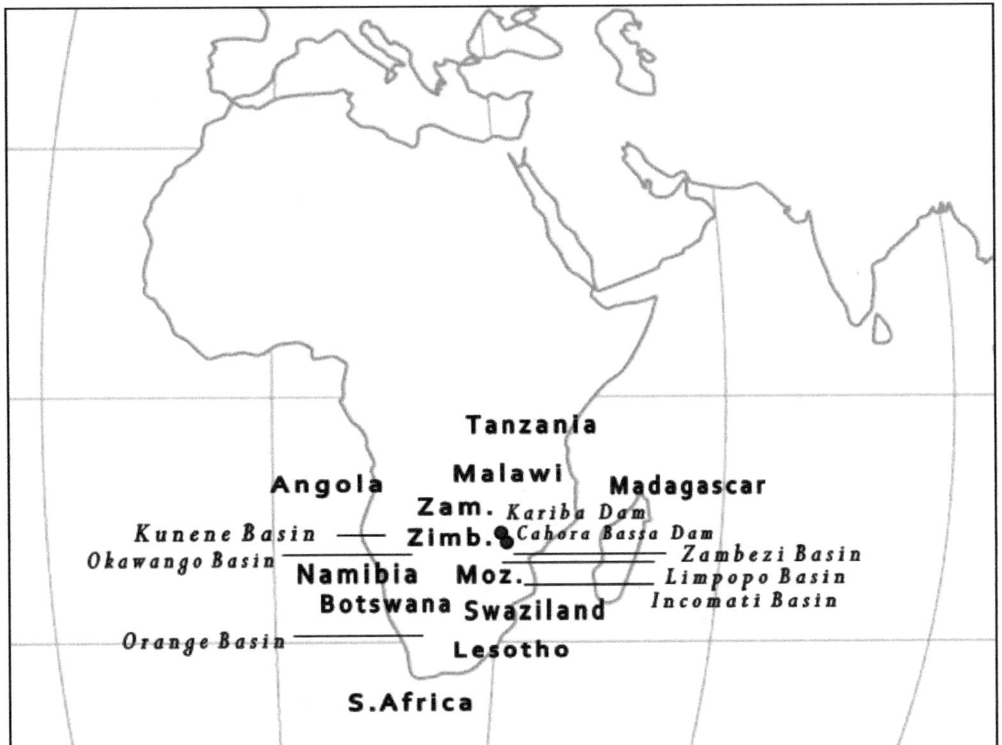

Hydropolitical risk as a tool for improving water resource management (Southern Africa)

Integrated water resources management in the Senegal River basin

Agathe Maupin*

From hydro-pessimism to hydro-optimism in Southern Africa: Hydropolitical risk as a tool for improving water resource management

Abstract. Is it possible to speak about risk without being a pessimist? The following paper aims at identifying a specific sort of risk linked to water resource management, in Southern Africa, identified by the author as "hydropolitical risk". The following analysis of hydropolitical risk as a tool aims to build a bridge between two tendencies in the hydropolitics research field: the first one concerns the *Hydropolitical Southern African Complex* (Turton, 2002), which can be seen as an optimistic approach to Water Resource Management (WRM), in opposition to the classification of Southern Africa as a "Basin at Risk" (Wolf, 2003), which can be perceived as a pessimistic point of view. The strong conceptualization of risk, especially in French, German and English-speaking geographical and political studies, will complete the definition of hydropolitical risk, which will then be applied to Mozambique as a case study.

Determining how hydropolitical risk is intensified by an obvious lack of connection between water management institutions from the regional to the local level may encourage changes in Africa's water management perspectives. Dealing with water scarcity, political instability, conflict potential in Southern Africa due to mismanagement practices, whilst taking into account the hydropolitical risk, may help to further develop innovations and alternatives in WRM in Southern Africa. Examples from various countries in the region will serve as data, with a particular focus on the Mozambican case study, by comparing the degree of regional water management's integration in former and new national water policies and its impact on local practice.

* Doctor in Geography. Associated Researcher, UMR 5115. Laboratoire LAM, Les Afriques dans le Monde, Bordeaux, France; agathe.maupin@gmail.com.

Keywords. Southern African, Hydropolitical Risk, vulnerability, water management, water policy

Introduction

Hydrological risks are easily identified when hazards like droughts and floods are challenging societies' vulnerability and security: they are all well-recognized water-related risks. The political risks are more connected to the actions of states; they are defined as risks without any border, the mismanagement of which has a direct impact on resources and societies. They might drive states into conflicts, especially when cooperation does not exist or is ineffective. Regarding the importance of water resources in the development and the resilience of a society, both hydrological and political risks should be taken into account in the elaboration of Water Resource Management (WRM). Hydropolitical risk, which was first used as an indicator to characterize States in Southern Africa (Turton, 1999), will be explored here as a tool for improving WRM by linking its various elements.

After reviewing some of the main contributions in risk studies debating conflicts and cooperation, the three dimensions of hydropolitical risk need to be explored within the context of change in the development of river-basins in Southern Africa. What, if anything, can the Mozambican case tell us about the hydropolitical risk contained in WRM? Indeed, the Mozambican government has recently renewed its efforts to implement river-basin management units and river-basin commissions over transboundary rivers. To address this question, it is important to argue here that there are multiple approaches (not only the state-centric one) to consider. The concept of hydropolitical risk will be further developed in an attempt to determine these new approaches including various actors, and an unexplored *time-lag* (Wolf, 2007) between the state-centric dominant overview in WRM and its implementation at the river-basin level. The main purpose remains to improve water management.

Theorizing Hydropolitical Risk: a risky enterprise?

Where does hydropolitical risk come from? To conceptualize the idea of hydropolitical risk, trends and debates in French, German and English-language research has to be considered in the geographical and political fields.

Declaring divorce from the nature/culture, hazards/vulnerability dichotomies

Over recent decades, debates and literature concerning risk and governance, in addition to Integrated Water Resource Management (IWRM) have proliferated, largely among experts, scholars, politicians, etc. Risk constitutes a large research field, particularly regarding water issues with various studies of floods, scarcity, cyclones, etc. At the risk of stating the obvious, the core of the analysis of water-related risk is based on the normative vision that risks will produce conflicts, leading us onto the hydro-pessimistic path.

As a starting point, one should consider that the same dichotomies between *nature* and *culture*, *hazards* and *vulnerability* can be found in French, Anglo-Saxon and Germanic definitions of "risk". The couple *natural hazard* and *vulnerability*, with the *natural hazard* referring to "a natural uncontrollable physical event" (Frampton, 1996) and the *vulnerability* referring to the passivity of the society facing the natural events are part of the background to risk studies. However, this concept of risk was called into question early. As compared to the academic definition "we consider risk to human communities from spontaneously occurring geophysical events such as storms, floods and unseasonable cold or drought" (Hewitt, 1983). The following point of view was a first step towards a non-deterministic approach: "the above concepts emphasize the natural aspect of natural hazards. Disasters can also be viewed from a sociological or a humanistic point of view. [...] by examining the human response to natural hazards at the personal or group level" (Bryant, 1991).

Twenty years ago, authors were already pointing out the need for a more systemic approach to risk studies, considering connections and interactions between resources and societies, and no longer the equation of hazards facing vulnerability as the previous definition of risk had done (Blaikie, 1994). Nowadays, researchers tend to pay more attention to the interaction between the elements, to define risk as "the product of an interaction of physical and human forces that, in combination, determine the significance and impact of disasters" (Tobin and Montz, 1997). Even nature can be perceived and thought of as a social construct: "I argue that in effect the environment is itself a social construction" (Cannon, 1994). These evolutions in the approach to risk have led to the conclusion that risk is itself perceived of and built as a social construction (Beck, 1992).

Hydropolitics and risk: a successful remarriage?

Since the perception of risk has changed from a hazard-centred approach to a society-centred approach, the field of risk management has developed accordingly. The focus of concern is, therefore, the way societies cope with risk, and, in the particular case of water-related risk, political scientists have made their point by developing a management framework called hydropolitics (Waterbury, 1979). Hydropolitics is defined within the field of international relations. It emphasizes "the systematic examination of the interaction between states, non-state actors and individuals, within the national and international domain as regards the authoritative allocation and/or use of international and national water resources such as rivers, aquifers, lakes, glaciers and wetlands" (Meissner, 2000). The hydropolitical approach of WRM is strongly based in the field of water policy, questioning power issues. One dimension of power is of special relevance in WRM, described as

> the 'third dimension', referring specifically to the "power to prevent people, to whatever degree, from having grievances by shaping their perceptions, cognitions and preferences (Lukes) [...] this third power dimension [is referred to] as 'ideational power'". The State shows his leadership intentions by giving his way of managing Water Resources as the best (and so on unique...) solution, without introducing any other alternative way to manage Water Resources (Zeitoun and Allan, 2008).

The *London Water Research Group* (LWRG)[1] currently focuses on transcending the dichotomy of conflict/cooperation (those of nature/culture and hazards/vulnerability mentioned above remain to be tackled) it remains to the nature/culture and then to the hazards/vulnerability couples analyzed above) to improve transboundary WRM.

The *Transboundary Water Interaction Nexus Model* (TWINS) provides a three dimensional perspective of WRM, including conflict, cooperation and political economy (Zeitoun and Mirumachi, 2008). The power question seems of great interest to the extent to which it highlights how states are using and could use further WRM to express their leadership inside and outside of their borders. The concept of *hydrohegemony* (Zeitoun and Allan, 2008) describes the influence exerted by the stronger state, mixing coercion (force) and authority (consent), to maintain its leadership without provoking its neighbour states into open conflict. Even when the *hydrohegemon* (meaning the stronger state) proposes solutions somehow issued from the international community, like IWRM, and the use of the river basin as a management unit, it is still a 'who gets to say what' perspective, which implies the closure of alternatives, especially for the other riparian states (Zeitoun, 2008). Hydropolitical risk could therefore be identified as a major risk originating in the lack of efficiency, or lack of alternatives, of political choices in WRM. However, it is important to keep in mind that hydropolitical conflicts remain potent and that they may occur in future cases when economies are weak and unable to adapt to their specific water situation. In the *Southern Africa Hydropolitical Complex*, developed to highlighting the regional benefits of South African leadership in regional WRM, both the necessity and the flexibility of a leadership position in a given river-basin are clearly stated (Turton, 2002). Regarding the recent development of conceptual schemes and theories in WRM mentioned above, it seems important to define precisely the hydropolitical risk as a social one tightly connected to WRM practices.

As water becomes more and more highly politicized, even securitized resource, because its management relies on a state-centred water-policy,

1 The LWRG depends on the KCL and SOAS and this research group has been implemented at London to regroup researchers from various countries around the world, all working in the hydropolitics field.

WRM should be considered as the main factor likely to increase or decrease the level of hydropolitical risk.

When hydropolitical risk meets IWRM in Southern Africa

The 1992 Dublin Principles introduced what is now known as Integrated Water Resources Management (IWRM). It brought recognition of the finite nature of water together with more holistic approaches, setting the goal of considering economic and social dimensions and making water management more participatory.

This inclusive approach is based on the perception of water as an integral part of the ecosystem, a natural resource and a social and economic good, the quantity and quality of which determine the nature of its utilization, (United Nations, 1992). IWRM was intended to be implemented at a river basin level. The trend behind of the IWRM concept has translated into a multitude of attempts to establish River Basin Organizations (RBOs) of all sorts, which are believed to be necessary mediums for achieving the integration of water managers, users, and ecosystems in global partnership. Southern Africa, which is characterized by multiple shared watercourses and aquifers, has started to implement the so-called RBOs with more or less success; this point will be developed further in this paper. However, what should be pointed out here is that IWRM is also a political and even ideological construct (one could even say product). It draws its strength from its systemic approach, but since it remains a model for managing water resources, and since it is a governance model, IWRM is likely to face increasing difficulties in its implementation: "*How interconnected and nested waterscapes can be managed by discontinuous nested political/administrative and social levels remains a fundamental question fuelling an endless search for elusive governance systems that unite nature and society*" (Molle, 2006). Indeed, the political temporalities are somehow shifted backward the societal and environmental ones.

The Southern African "hydropolitical bet" or the likelihood of the hydropolitical risk occurring in Southern Africa

The last aspect of hydropolitical risk comes from a concrete application of the *Southern African Hydropolitical Complex* (SAHPC) (Turton, 2002), in reaction to the classification of Southern Africa as a "basin at risk" (Wolf, 2003).

Hydropolitical risk has been identified through fieldwork in Southern Africa. Indeed, large portions of Southern Africa are arid, poor, sparsely populated, affected by conflicts over various resources, which used to be part of the criteria for classifying Southern African river-basins in the "at risk" category.[2] Six basins were identified as "at risk", namely the Cunene, the Incomati, the Limpopo, the Okavango, the Orange and the Zambezi basins (Wolf et al., 2003). But neither aridity nor scarcity have yet seemed sufficient to drive states into a conflict over water resources.

Hydropolitical risk has therefore been identified *a priori* in Southern Africa, as the result of the cumulating factors mentioned above. However, these factors still need to be clarified and linked together in order to prove a lack of connection between the various levels of WRM. After this brief overview of the conceptual background of hydropolitical risk, the next step will be to test it in Southern Africa through the Mozambican case study, in order to challenge it as an efficient tool for improving WRM.

The Southern African framework

Why is Southern Africa a relevant choice to identify and conceptualize the concept of hydropolitical risk?

The first reason lies in the SAHPC, which was elaborated to show the link between river-basins and states (Turton, 2002). Based on the *Security Complex Theory* (Buzan, 1991), it is an assessment of Southern African riparian states' relationships regarding the international river-basins

2 Aaron Wolf and his team at the University of Oregon elaborated in 2003 a Transboundary Freshwater Dispute Database (TFDD) available online: <www.transboundarywaters.orst.edu> (accessed 2007–2009).

they share. The SAHPC focuses on the distinction between river-basins and riparian states, using the following terminology of "pivotal" and "impacted" States to qualify the role of these states inside the Southern African transboundary river-basins. The main conclusions that can be drawn from this pattern are that all river-basins are not equal in terms of strategic importance, and most importantly, that riparian states do not hold the same power and access over the international river-basins crossing their territories, which refers directly to the hydrohegemony theory mentioned above.

The second reason was established before the SAHPC design. The degree of hydropolitical risk has already been evaluated in Southern Africa to introduce the importance of the "adaptive capacity" of a society facing water scarcity, according to its development level (Turton, 1999).

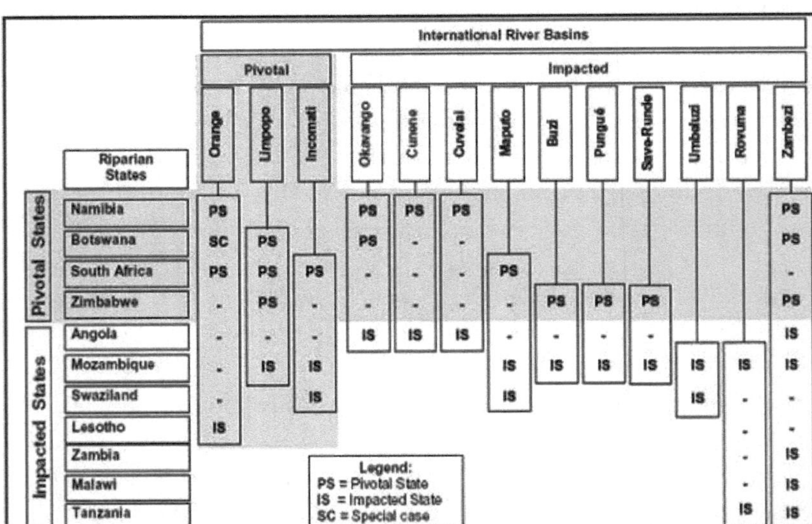

Figure1. The Southern African Hydropolitical Complex (Turton, 2002).

All the nine States sharing international rivers in Southern Africa have been characterized according to their *Hydropolitical Risk Profile* (HPRP). The purpose was to underline the link between water poverty and water scarcity: Malawi, Mozambique, Tanzania, South Africa, Zimbabwe have been evaluated as States with a high HPRP, (Swaziland and Lesotho were not mentioned but can be considered as belonging to this category as well), compared to Angola and Zambia, which have been classified as States with a low HPRP, while Botswana and Namibia have been considered having medium HPRP. The criteria for establish the distinction were poverty, scarcity and economy. According to these criteria, water-policy interventions within states combining a weak economy and poverty without an abundance of resources (Malawi, Tanzania and Mozambique) would have limited means to rescue them from ecological catastrophe. The main conclusions drawn were that scarcity concerns not only natural resources but also social and economical resources: "there are major gaps in the knowledge regarding the social, institutional and hydropolitical dynamics of water sector reform. It is unknown, for example, exactly how reform is generated and how policy is actually made or impacted on by various actors in the water sector" (Turton, 1999).

Hydropolitical riskc annot be limited to the study of the political risk of conflicts about water resources when river-basins are shared by several states. It also results from water resource mismanagement emerging through water-politics, which could therefore lead to unsustainable development and environmental disasters, the previously so-called ecological catastrophe.

However, given all the changes that have occurred in recent decades, especially the development of international Laws over water resources, the increase of environmental considerations in new water policies in Southern Africa, how would the situation be evaluated now? Ten years later, the criteria used to define the HPRP might have changed slightly.

3 SADC: the Southern African Development Community is a regional organization including fifteen members States (Angola, Botswana, Lesotho, Madagascar since 2005, Malawi, Maurice, Mozambique, Namibia, The Republic of South Africa, the Democratic Republic of Congo, Seychelles, Swaziland, Tanzania, Zambia and Zimbabwe).

The Southern African case is interesting due to the existence of the Southern African Development Community[3] (SADC), which is trying to improve the management of resources, in order to reach its regional development goals. The fifteen SADC members States (except the Democratic Republic of Congo) signed in 2000 the SADC *Revised Protocol on Shared Watercourse Systems*, which led them to implement international RBOs when this Water Protocol entered into force in 2003. Indeed, relying on the 1997 UN Watercourses Convention, this Water Protocol recommends the river-basin unit as the best one for managing water resources.

The point is that water flows do not respect administrative boundaries. Adopting a river-basin approach seems logical, especially for developing co-operative units between and within states. Nevertheless, it can be argued that even if a river-basin can be defined by the watershed limit of a system of waters both ground and surface flowing to a common terminus, the River Basin Management (RBM) Unit still does not appear to integrate this system of waters within its broader natural environment and its social, economic, and political contexts; furthermore WRM is implemented by states. So far, the water institutions, even when they are declared independent from states' national interests, still have to account for their actions to higher governmental authorities.

In South Africa, the RBM Units, known as Catchment Management Agencies (CMA), were supposed to establish the "rules of the game" and mechanisms for governing these interactions between water users and environmental needs at the river-basin level. Actually, ten years after the launch of the new Water Policy in South Africa (based on the new National Water Act, 1998) only three RBMs are in place. Botswana is actually on the way to water management modernization with a new Water Law project, presented in 2006 by way of a National Resource Strategy Management Plan, following that of South Africa. The same occurred in Namibia in 2004. In Zimbabwe, the Water Act and the Zimbabwe National Water Authority Act (ZINWA) Act of 1998 were also supposed to divide the country in seven catchments based on hydrological boundaries. In ten years, these SADC member states have all chosen to rely on river-basin hydrological boundaries, in order to decentralize their water resource management. However, they are all facing difficulties in implementing their new policy.

Regarding the restructuration of Water Resource Management in SADC members states, it seems necessary to consider the question of management scales, which is of particular relevance to the easily identifiable disconnections between regional, national and local water resource management levels. Hydropolitical risk also directly rises from the obvious lack of connection between Water Management Institutions from the regional to the local level, a disconnection which is intensified by the focus on the river basin as the only management unit approach. The example of Mozambique will be used to highlight these disconnections and to identify hydropolitical risk.

The Mozambican case: when water-politics meet development

The SADC member states share scarce water resources which are unevenly distributed both in time and space. Most states regularly experience severe droughts and damaging floods, and especially Mozambique (the latest floods occurred in January 2010, submerging Maputo's buildings' feet in several inches of water). It is important to first have a look at the physical characteristics of water resources in Mozambique, especially regarding the geographical situation of this state as the downstream riparian for all its shared river basins. Indeed, Mozambique relies on water resources that are generated outside its own territories for 54% (DNA, 1999). Three of the six basins identified as "at risk" (Wolf et al., 2003) are shared by Mozambique with more than one state, namely the Incomati, the Limpopo and the Zambezi. Most of Mozambique's rivers run from west to east, draining water from the Highveld to the plains along the coastal area to the Indian Ocean; these rivers are characterized by reduced runoff coefficients. In Mozambique, the availability per capita of surface water resources is around 5500 cubic meters/inhabitant/year, taking into account only the runoff generated inside the borders (when including the flows from neighbouring states, these figures reach around 11 000 m3/inhabitant/year). Furthermore, cyclones occur relatively frequently, despite the natural protection afforded by the island of Madagascar. Over the past ten years, several cyclones have caused major destruction and loss of human life, such as Nadia (1994), Connie, Eline, Gloria (2000), Favio (2007), Jokwe (2008). By comparing these figures with the water availability in other parts of the world, Mozambique has no shortage of water, but the situation is worsened by the geographical and seasonal distribution of the water, population growth, the declines in flow from

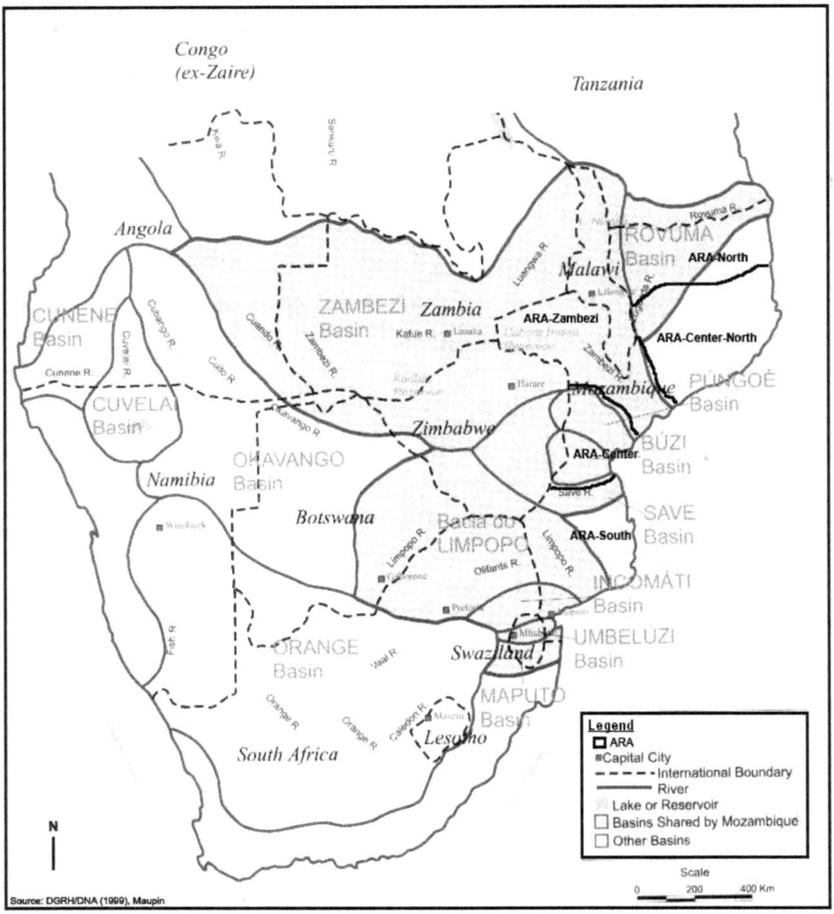

Figure 2. The Mozambican hydropolitical position in Southern Africa.

upstream riparian States (resulting in periodically severe droughts in the Mozambican provinces of Gaza, Inhambane, Maputo, Tete).

In 1991, Mozambique elaborated a new Water Law (*Lei de Águas*), followed in 1995 and in 2007 by a new National Water Policy (*Política de Águas*). They all focused on bringing together beneficiaries of water, stakeholders, and local state administrations. They divided the country into five distinct areas, called *Administração Regional de Aguas* (ARA), which are regional water administrations. The ARA administrative boundaries correspond to a river basin like the ARA-Zambezi, or to a wider area, like the ARA-South. These five regional water administrations[4] are still coordinated by a national one, the *Direcção Nacional da Águas* (DNA) which is in charge of applying the Mozambican National Water Policy. In 2007, Mozambique elaborated another Water Policy, by relying on a Water Resources Management National Strategy (Estratégia Nacional de Gestão de Recursos Hidricos, 2006) to complement the former Water Policy of 1995, by including the legacy of the 2000 SADC Revised Protocol mentioned above. The ARA are structured in river-basin management units, for example, ARA-South is divided into four Units, namely the Umbeluzi River Basin Management Unit (UGBU) that includes the Maputo river-basin, the Incomati River Basin Management Unit (UGBI), the Limpopo River Basin Management Unit (UGBL), the Save River Basin Management Unit (UGBS). The ARAs were established in order to decentralize integrated water resources management in Mozambique. However, these regional water administrations are still reporting to the DNA for their actions at the river-basin level.

If one briefly considers the Mozambican case, major events occurred frequently, such recurrent floods in 2000–2001, 2007–2008 and droughts in 1992, 2002. These natural events, combined with the high vulnerability of a developing country like Mozambique, caused the death of thousands

4 ARA-South (Sul) was established in 1993, ARA-Centre (Centro) in 2004, ARA-Zambezi (Zambeze) in 2005, ARA-Centre North (Centro-Norte) in 2006, ARA-North (Norte) in 2007 (according to the published public announcement in the Official Government Gazette).

5 The Mozambican Growth Domestic Product rate around 8% per year on average since 1996 was reduced to less than 2% following the 2000 floods events according to the World Bank and IMF data online.

people (Filipe, 2003), the displacement of hundreds of thousands others (United Nations) and the decrease in an already low Growth Domestic Product.[5] These disasters are the result of a combination of multiple factors. If hydropolitics globally focuses on conflict management over international waters, hydropolitical risk, used here as an analytical tool, provides another approach to identifying global mismanagement, where "the rate of change within the basin exceeds the institutional capacity to absorb that change" (Wolf, 2007). Hydropolitical risk highlights that the core of an integrated water resources management is neither hazards, nor society. It is both, and more, specifically the (mis)management of resources by societies.

To manage specific disasters and avoid their repetition, hazard management (especially in cyclone-prone areas) is needed for river basin management (especially when river basins are shared between several states) which would include alert systems and data exchange between upstream and downstream riparian states (for example, the 2007 floods in the Zambezi Province in Mozambique were exacerbated by the need to increase the discharge from the Kariba and the Cahora Bassa Dams).[6] So, what does it take to manage hydropolitical risk? Despite the 2000 SADC Revised Protocol and the ratification of International Treaties establishing river-basin commissions at an international level, there is still a tendency among the upstream states of an international river to treat the basins of the tributaries of the shared watercourse as independent of the river basin. Consequently, the upstream state believes it is its sovereign right to manage it solely (concerning the Zambezi Watercourse Commission, the

6 Notícias: <www.jornalnoticias.co.mz/pls/notimz2/getxml/pt/contents/25772/20070205> (Journal of Mozambique accessed, June, 12th, 2009).

7 An Inception Study on Strategic Interventions for Negotiations on International Rivers, ordered by the MOPH-DNA-GRI, clearly expresses the Mozambican interest in international treaties: "Equitable and reasonable utilisation of shared watercourses » is important on all basins, but especially the Zambezi, where Zambia has actively attempted to lead this process. Mozambique requires a clear and strategic approach to identifying and putting forward « all relevant factors » in a way that ensures it receive its « equitable and reasonable utilisation » of the use of its waters" (MOPH-DNA-GRI, 2004).

Zamcom agreement on the Zambezi was signed in 2004, but several states have yet to ratify it, partly because they consider some tributaries of the Zambezi under their territorial sovereignty).[7] Political risks are increased by disconnections between management levels and the exclusion of actors, mainly civil society, especially in state-centric approach. At the moment, most of the cooperative multilateral agreements are limited to regular meetings of the river-basin commission, with a focus on the exchange of data and development plans, without a complete joint operating infrastructure. Despite the signature of the 2003 Treaty, the Limpopo Watercourse Commission (Limcom) was just implemented in Mozambique, the Limcom Secretariat host country, in January 2009. If one takes a closer look at the Zamcom and Limcom situations, it becomes obvious that States are the principle actors in the implementation, development and outcomes of these commissions. Finally, the "risk society" emerged from its lack of adaptive capacity, which resulted in a decrease in societies' resilience. The purpose of this hydropolitical risk analysis is to highlight the need to take into account the interactions in all these water spheres, in order to improve risk management. To summarize the hydropolitical risk identified through the specific case of the SAHPC, it might be that this risk exists not only when threat of conflict over water resources exists, but also as soon as a society faces difficulties in its water resource management. Such difficulties are raised by various factors, ranging from environmental resources to the capacity of the society to cope with change. Hydropolitical risk is therefore increased or decreased by the efficiency of institutions, which itself relies on the degree of cooperation between the actors sharing water resources, and the degree of development of the society in question.

Improving water management = decreasing hydropolitical risk? Not such a simple equation to solve

WRM appears to be itself at the core of hydropolitical risk factors, especially when it is set up only at the national level and is seen, therefore, as an exclusively political issue, driven merely by "state-will". As the concept of hydropolitical risk has highlighted, the system of risk management relies mainly on institutional capacity to absorb any change in the system. The RBOs implemented at various levels, from international to local, have great difficulties in adapting to rapid change, partly because they remain state institutions. They lack flexibility and, as they are dependent on states, they rely on political agendas, which often results in a "time lag" (Wolf, 2007) between the time they face a problem, define their common interests and then elaborate a win-win scenario and sign agreements. But by the time they eventually agree, the impacts (on populations, ecosystems…) have already occurred.

Questioning the river-basin as a management unit

Do river basins have the institutional capacity to mitigate the states' sovereign interests over water resources? Looking at the implementation both of river basin commissions at the international level and river basin management units at the local level, the *hydrosocial contract* (Turton et al., 2007) falls far short of fulfilling its obligations, due to the incapacity of governments to include marginalized segments of the population. If the IWRM approach has contributed to developing more inclusive and resilient institutions, especially through decentralization processes, it has also created an institutional "melee" (Nicol and Mtisi, 2003) that has increased the confusion over roles and responsibilities between institutions, managers, local actors, and stakeholders. The result is that the multiplication of RBOs has pushed water local needs into the background, sometimes by "disconnecting" people. The "users-pay" principle, included in the IWRM framework, has contributed to intensifying the debate between consumers and providers. The Phiri, Soweto, water riots in South Africa, in 2003, arose as the result of protests against the installation of pre-paid water metres by the Johannesburg Water and the City of Johannesburg. They resulted from the tension between supporters of "water as an economic good" and "water as a human right". This brings us to the conclusion that: "Like other buzzwords, IWRM is likely to fall prey to opportunistic hijacking by interest groups". It may become "no more than bureaucratic hoops to jump through in order to obtain financial aid" (Chéret, 2004). Alternatively, it be used by consultant firms to repackage their business-as-usual strategies (Molle, 2005) or to "obscure the top-down nature of projects, making them appear to be bottom-up approaches so as to dampen criticism and attract Western economic and technical assistance" (Tsering, 2005, on China) (Molle, 2006).

So perhaps the time is right to think of another basis for WRM. Here comes the main challenge: how to determine the best way to integrate all the dimensions that were developed above and to shape them in an equitable and sustainable way? The question of which role states should play in WRM remains a highly politicized and closed one. As the hydropolitical risk comes mainly from disconnections between WRM levels, leading to misunderstandings regarding water demands, water offers, and water perceptions, identifying it is a way to identify what the problems are in the IWRM framework, namely the lack of choice and alternatives. Improving the management of water resources presupposes a better identification of the rights, needs and duties of all water-carriers. This realization conversely highlights the potential damage that might result from a lack of coordination between regional and national levels and a lack of efficiency on the part of the water institutions (and states).

A long walk to civil society?

The study of the Mozambican case reveals that hydropolitical risk mainly lies in the disconnections between the main institutions managing water resources and their inefficiency to promote development and protect the interests of all the river-basin actors. At a regional level, river-basin commissions may one day provide a useful mechanism for managing international water resources, which will involve a stronger political will and the real inclusion of civil society in river-basin management. Indeed, the river-basin approach appears strongly linked to a state-centric conception of WRM, In Mozambique, despite the process of decentralization, the government and the IWRM approach is still behind any water initiatives. WRM and IWRM are still perceived and managed as part of states' power and prerogative. One still has to realise that remains "an almost instantaneous, unquestioned and perhaps unconscious acceptance of the state as the primary organisational node for water resources management" (Swatuk and Wirkus, 2009).

However, progress in the interconnections between basin-wide institutions and the in-country catchment organizations has already been achieved. The ARA-Zambezi in Mozambique is a key institution, dealing directly with its equivalent in Zimbabwe. In this case, several water forums, where water users are included in the planning issues, have already taken place. If all the actors concerned with water resources have not yet being consulted at all management levels in the same river-basin, "linking catchment-based forums across borders to form basin-wide platforms could deepen and balance interstate agreements, enhance mutual understanding, and widen the perspective that water users have of their basin" (Van der Zaag, 2009; Böge, 2006). In *Democratizing river basins* Sneddon and Fox (2008) include stakeholders and water users in WRM, which appears to be the precondition for a less risky, and, so to speak, a more sustainable WRM.

Conclusion

Taking all water resource management levels in Southern Africa into account is a difficult task, which implies juggling a collection of water institutions, including numerous actors, users and goals. Southern Africa, with its particular socio-political landscape, history, environment and transboundary rivers, is a place from which innovations and alternatives in WRM might arise, by confronting the difficulties the state-centric approach faces in the river-basin unit. The *territorial trap* (Furlong, 2006) could close up around Southern Africa WRM or WRM could be conceptualised differently.

Identifying the factors increasing and decreasing hydropolitical risk is a contribution to the greater understanding of the dynamics at stake in this area through the building of a "new water architecture" (Swatuk, 2002), especially by bringing the lack of connections between WRMI into the foreground. But the main difficulty remains in the lack of connections between the political importance attached at the national level by the states, mainly to modify water law, and the lack of their political commitment and investment at the level of implementation, and, eventually, at the local level:

> The level of investment in water management in the SADC region varies from country to country but in general it is not very high. There is a clear need to improve the level of investment particularly in respect to information and data collection, infrastructure development, operations and maintenance of infrastructure once developed. There is also a need to invest in capacity building and institutional strengthening of those institutions responsible for water resources management and development (Phera Ramoeli, 2008, Head of the SADC Water Division).

References

Beck, Ulrich (1992). *Risk Society: Towards a New Modernity*. London: SAGE.

Blaikie, P. et al. (1994). *At risk-natural hazards, people's vulnerability and disasters*. London: Routledge.

Böge, V. (2006). *Water governance in Southern Africa-cooperation and conflict prevention in transboundary river basins*, Bonn: Bonn International Center for Conversion (Brief 33).

Boletim da República. No 34/95. Resolução N° 7/95. Conselho de Ministros: Polícia Nacional de Águas, Moçambique.

Bryant, E. (1991). *Natural hazards*. Cambridge: Cambridge University Press.

Buzan, Barry (1991). *People, States and Fear: An Agenda For International Security Studies in the Post-Cold War Era*, 2nd Edition. Hertfordshire: Harvester Wheatsheaf.

Cannon, T. (1994). Vulnerability analysis and the explanation of "natural" disasters. In Varley, A. (ed.), *Disasters, development and environment*. London: J. Wiley and Sons.

DNA (1999). *Water Resources of Mozambique Synopsis*. Maputo.

Filipe, Jossias (2003). *Moçambique 2000 – As Águas da Morte*. Moçambique Editora, Lda.

Frampton, S. et al. (1996). *Natural hazards: causes, consequences and management*. London: Hodder et Stoughton Educational.

Furlong, K. (2006). Hidden Theories, Troubled Waters: International Relations, the 'Territorial Trap', and the Southern African Development Community's Transboundary Waters. *Political Geography* 25 (4): 438–458.

Hewitt, K. (1983). *Interpretations of calamity from the viewpoint of human ecology*. London: Allen and Unwin.

Hewitt, K. (1997). *Regions of risk: geographical introduction to disasters*. London: Longman.

Meissner, R. (2000). Hydropolitical hotspots in Southern Africa: Will there be a water war? The case of the Kunene river. In H. Solomon and A. Turton (eds). *Water*

Wars: Enduring Myth or Impending Reality? p. 103–131. Africa Dialogue Monograph Series 2.

Molle, François (2006). *Planning and managing water resources at the river-basin level: Emergence and evolution of a concept.* Colombo, Sri Lanka: International Water Management Institute, IWMI Comprehensive Assessment Research Report 16.

Nicol, A.; Mtisi, S. (2003). *The Politics of Water: a Southern African example.* Brighton: Institute of Development Studies. Sustainable Livelihoods in Southern Africa Research Paper 20.

SADC – Southern African Development Community (2000). *Revised Protocol on Shared Watercourses*, <www.sadc.int/index/print/page/159> (accessed 2009–05–14).

Sneddon, C.; Fox, C. (2008). River-basin politics and the rise of ecological and transnational democracy in Southeast Asia and Southern Africa. *Water Alternatives* 1 (1): 66–88.

Swatuk, L.A. (2002). The New Water Architecture in Southern Africa: reflections on current trends in the light of 'Rio +10'. *International Affairs* 78 (3): 507–530.

Swatuk, L.A.; Wirkus, L. (eds) (2009). *Transboundary water governance in southern Africa: Examining underexplored dimensions.* Baden-Baden: Nomos Verlagsgesellschaft.

Tobin, G. A.; Montz, B. E. (1997). *Natural hazards-explanation and integration.* London: Guilford Press.

Turton, A. R.; Hattingh, H. J.; Maree, G. A.; Roux, D. J.; Claassen, M. and Strydom, W. F. (eds) (2007). *Governance as a trialogue: Government-society-science in transition.* Pretoria and Berlin: CSIR and Springer.

Turton, A. R. (1999). *Water scarcity and social adaptive capacity: towards an understanding of the social dynamics of water demand management in developing countries.* London; SOAS, MEWREW Occasional Paper 9.

Turton, A. R. (2002). *Hydropolitics in the developing world. A Southern African perspective.* Pretoria: AWIRU.

Turton, A. R. (2005). *A Critical Assessment of the Basins at Risk in the Southern African Hydropolitical Complex.* Pretoria: CSIR Report.

Van der Zaag, P. (2009). Southern Africa: Evolving Regional Water Law and Politics. In Dellapenna, J. W. and Gupta, J. (eds), *The Evolution of the Law and Politics of Water.* Springer Science and Business Media BV.

Water analysis, *Water Policy* 10 (2): 3–12.

Waterbury, John (1979). *Hydropolitics in the Nile Valley.* New-York: Syracuse University Press.

Wolf, Aaron (2007). *A long term view of Water and Security.* International Waters, National Issues, and Regional Tensions, External expertise for the WBGU. Berlin: WBGU.

Wolf, A. T.; Yoffe S. B.; Giordano, M. (2003), International Waters: Identifying basins at risk. *Water Policy* 5: 29–60.

Zeitoun, Mark; Allan, J. A. (2008). Applying hegemony and power theory to transboundary water analysis. *Water Policy* 10: 3–12.

Zeitoun, Mark; Mirumachi, Naho (2008). Transboundary Water Interaction I: Reconsidering Conflict and Cooperation. *International Environmental Agreements* 8 (4): 297–316.

Ibrahima Anne*

Integrated water resources management in the Senegal River basin

Abstract. The integrated development program of the Senegal River basin, initiated various decades ago through the *Organisation Pour la Mise en Valeur du Fleuve Sénégal* (OMVS), has had tremendous impact on the livelihood of the populations of member states.
It is a program of great regional importance in this part of western Africa, often pointed out as an example of regional cooperation for transboundary river management.

However, the populations living by the banks of this river, which in reality has always served more as a link than a boundary, were those who suffered the most from the changes induced by this infrastructure program.

It is thus of paramount importance to monitor and study the quantitative and qualitative modifications introduced by the program in the different aspects of the life of the populations. It is also important to point out the positive aspects that can serve as examples for future progress, as well as the solutions considered necessary for the resolution of the negative impacts created by the implementation of the program.
Keywords. Transboundary river basin management, dam construction influences, legal framework, governance.

* Centro de Estudos Africanos da Universidade do Porto (CEAUP); Degrémont SA, Rua General Ferreira Martins nº8, 3ºA/C/D, Miraflores, 1495–137 Algés, Portugal. anne.ibrahima@gmail.com.

The Senegal River basin

This is a 1,800 km lifeline for a multi-ethnic, multicultural population, that runs through sub-Saharan Africa in a predominantly desert region. It is the second largest river in West Africa. The region is included geographically in the sahelian sudanese zone, which is subject to chronic droughts and famine and where water availability represents the most important stress for the population (Reizer, 1986).

Population figures have been updated based on growth rates in each country. Irrigation is the motor of development of the basin, especially in the valley and the delta, and livestock has always been a major activity. Following agriculture, fishing is the second largest economic activity of the basin <www.omvs.org>.

As seen, almost 15% of the population of the three states lives in the basin area, and the vast majority of irrigated land in Senegal and Mauritania are located in the river basin.

A difficult context

The climate is tropical with two seasons: one rainy (from June to October) and one dry the remainder of the year. The region is entirely covered by the intertropical front (FIT). Rainfall is between 800 mm and 1200 mm a year. One important characteristic of the climate is represented by the annual irregularity of rainfall (Reizer, 1986).

Before the dams were filled in the mid-1980s, the local inhabitants depended directly on rainfall (rain crops) or on floods (flood recession crops), in particular in the upper basin in Guinea (Fouta Djallon Mountains). However, the dramatic and steady drop in rainfall during the 1960s and the 1970s led to the degradation of almost the entire base of natural resources (soil erosion, the disappearance of vegetation, drying up of surface water, salinity 200 km upstream from the mouth of the river, descrease in groundwater level, degradation and disappearance of

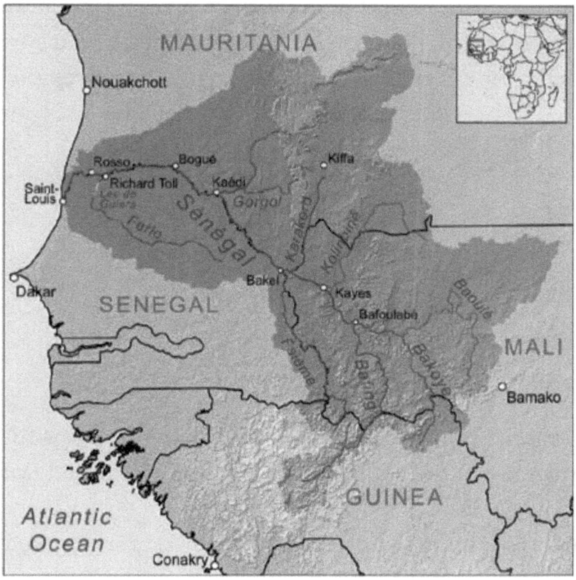

Figure 1. Senegal River Basin. Source: Prepared for the World Water Assessment Programme (WWAP) by AFDEC, 2002.

Zone Variable	Senegal river basin	Mali	Mauritania	Senegal
Population (million inhabitants)	3.5	11	3	10
Annual growth rate (%)	3	2.97	2.9	2.8
Urbanization rate (%)	NA	41	53	51
Farmland (ha)	823,000	NA	NA	NA
Irrigated land (ha) – national total	NA	76,630	49,200	71,400
Part in basin		4,000	44,449	67,830
Cattle (x1,000 units)	2,700	6,427	1,394	2,927
Sheep and goats (x1,000 units)	4,500	15,986	10,850	8,330
Fish catch (t/year)	26,000 to 47,000	100,000	620,000	395,000

Table 1: Some social and economic characteristics of the Senegal River Basin.

pastureland). Under these conditions, the local inhabitants could not produce enough to survive and the only alternative was emigration. Each year, a large percentage of the population, in particular young people, has left the valley and the delta for capital cities in the sub region (Abidjan, Bamako, Dakar, Libreville, Nouakchott, Brazzaville, etc.) or Europe (usually France or Italy) (Reizer, 1986).

The Organization for the Development of the Senegal River (OMVS)

Created in 1972 with a mandate to ensure food security and harmony among all riparian users, the main OMVS strategy was based upon several initiatives:

- construction two main dams and a series of dam infrastructures;
- energy;
- irrigatation;
- year-round navigation;
- original concept of "optimal distribution among users" rather than volumetric water withdrawals.

The main baseline for the organization lay in the recognition that river basin must be seen as the appropriate unit for management and development of water resources, especially in shared watercourses. Water resources are therefore considered an opportunity for cooperation and peace rather than a cause for conflict. The river basin approach is thus adopted as an implementation mechanism for regional projects (Observatoire de l'Environnement, 2006; Coyne et Bellier et al., 1996).

In fact, the river became a border due to colonial imposition. For riverside populations, especially between Mauritania and Senegal, the river even now is hardly seen as separating people, rather as joining them. By way of example, in the valley and delta, Mauritanian riverside populations (mainly from the *Halpular* ethnic group) still maintain some seasonal agricultural land on either side of the river, while the same is also true for the Senegalese riverside people. This issue of land possession is a crucial one and will have to be tackled by authorities of both countries, hopefully with an understanding of the historical and traditional conditions. Effectively, land has increased its economic importance with the perspectives of irrigation. A concomitant problem, mainly on the Mauritanian side, is that riverside lands has attracted greater attention from the agro-business sector, leading sometimes to situations of conflict with traditional community land ownership.

Legal and regulatory framework and governance

The first institutions to develop the Senegal River valley were created during the colonial period. On 25 July 1963, very soon after independence, Guinea, Mali, Mauritania and Senegal signed the Bamako Convention for the Development of the Senegal River Basin. This convention declared the Senegal River to be an "International River" and created an "Interstate Committee" to oversee its development. The Bamako Convention was supplemented by the Dakar Convention, signed on 7 February 1964, concerning the status of the Senegal River. The Interstate Committee laid the foundation for sub regional cooperation. On 26 May 1968, the Labbé Convention created the Organization of Boundary States of the Senegal River (OERS, *Organisation des Etats Riverains du Sénégal*) to replace the Interstate Committee, broadening the field of sub regional cooperation. Indeed, OERS objectives were not limited to the valorisation of the basin but aimed at the economic and political integration of its member states. After Guinea withdrew from the OERS, Mali, Mauritania and Senegal decided, in 1972, to set up the OMVS, which has pursued the same objectives (Reizer, 1977).

The OMVS has since created a flexible and functional legal framework enabling collaboration and co-management of the basin.

In 1998, the Programme for the Mitigation and Monitoring of Environmental Impacts drew attention to the lack of management and the risks involved. The Observatory of the Environment was created in

2000, as a tool for supporting basin management (GWP and INBO, 2009). That Institution was transformed into the Observatory of Sustainable Development in 2002, reflecting the political will of the governments to shift their focus from management of shared resources to the management of information and data for sustainable development.

The principal legal documents governing the OMVS are:

- The Convention concerning the status of the Senegal River (Convention relative au statut du fleuve Sénégal), 11 March 1972. According to the same, the Senegal River and its tributaries were declared an "International Watercourse", guaranteeing freedom of navigation and the equal treatment of users;
- The Convention creating the OMVS (Convention portant création de l'Organisation pour la Mise en Valeur du Fleuve Sénégal), 11 March 1972;
- The Convention concerning the Legal Status of Jointly-owned Structures (Convention relative au statut juridique des ouvrages communs), 12 December 1978, supplemented by the Convention concerning the Financing of Jointly-owned Structures (Convention relative aux financements des ouvrages communs, 12 March 1982.
- The Senegal River Water Charter, May 2002 (Charte des Eaux du Fleuve Sénégal) with the purpose of:

a) setting the principles and procedures for allocating water between the various use sectors;
b) defining procedures for the examination and acceptance of new water use projects;
c) determining regulations for environmental preservation and protection; and
d) defining the framework and procedures for water user participation in resource management decision-making processes.

These agreements declare that:

a) all structures are the joint and indivisible property of the member states;
b) each co-owner state has an individual right to an indivisible share and a collective right to the use and administration of the joint property;

c) the investment costs and operating expenses are distributed between the co-owner states on the basis of benefits, with each co-owner state drawing from the exploitation of structures. This distribution can be revised on a regular basis, depending on profits;
d) each co-owner state guarantees the repayment of loans extended to the OMVS for the construction of structures;
e) two entities manage the jointly-owned structures for the OMVS: one dedicated to the management and development of the Diama dam (SOGED, Société de gestion et d'exploitation du barrage de Diama), the other of the Manantali Dam (SOGEM, Société de gestion de l'énergie de Manantali), both created in 1997.

The OMVS functions with the following management bodies:

Permanent bodies:
a) Conference of heads of State and Government;
b) Council of Ministers;
c) High Commission, executive body;
d) Permanent Water Commission (CPE), made up of representatives of the organization's member states, and which define the principles of and procedures for the allotment of Senegal River water between member states and use sectors. The CPE advises the Council of Ministers.

The work of the CPE and the criteria used by the ministers for decision-making are based on the following general principles:
- reasonable and fair use of the river water
- obligation to preserve the basin's environment
- obligation to negotiate in cases of water use disagreement/conflict
- obligation of each riparian state to inform the others before undertaking any action or project that might affect water availability.

The objective of the OMVS method of water allocation is to ensure that local populations benefit fully from the resource, while ensuring the safety of the people and structures, respecting the fundamental human right to clean water and working towards the sustainable development of the Senegal River basin.

Non-permanent bodies:

a) an OMVS national coordination committee in each member state
b) local coordination committees
c) Regional Planning Committees (CRP)
d) Consultative Committee (CC)

The principal achievements of the Program are the following:

– year-round availability of freshwater in sufficient quantities (for agriculture, domestic uses, agro-industry, groundwater recharge), accompanied by reverse immigration of people who had left to find employment in the cities;

Figure 2. The Manatali dam and a view of the hydroelectric power plant.

- development of irrigated agriculture in the valley (with double cropping);
- partial opening up and stimulation of exchanges between areas where dams have been built and the rest of the sub region due to road construction (Reizer, 1977);
- access to health care for several villages near the dams with the construction of dispensaries and health clinics (Coyne et Bellier et al., 1996);
- access to drinking water installations for populations living near the dams;
- development of fishing activities for populations living near the Manantali dam, and immigration of highly skilled fishermen populations (mainly the Malian Bozo ethnic group) for the exploitation of the potential of the newly created reservoir (Anne et al., 1991, 1994);
- reappearance of local fauna and regeneration of vegetation;
- flow regulation to a decrease and elimination of flooding;
- cheaper electricity in the three member states thanks to the Manantali power plant.

On the other hand, the main negative effects of the program can be described as follows:

- as in the case of many large dam projects, displacement of populations living in the areas of dam construction (Manantali and Diama) occurred; for example, the number of displaced populations due to Manantali dam is estimated to have been 11,000 people (Reizer, 1977).
- water pollution caused by the development of irrigated agriculture and the agro-industry (CSS, SAED in Senegal, SONADER in Mauritania) (Reizer, 1986);
- proliferation of water weeds in the valley and delta, clogging water courses and contributing substantially to making the ecosystem more uniform;
- degradation of the fish population available for independent fishermen (quantity and quality);
- significant reduction of pasture land;

- riverbank erosion, especially in the upper basin where the topography is much more rugged;
- degradation of cultivated land;
- modification of the hydrodynamic characteristics of the estuary with the reduction of the "flushing" phenomenon (Reizer, 1977);
- disappearance of wetland areas; and
- last but not least, the proliferation of water-borne diseases (bilharzia [schistosomiasis], malaria, diarrhoea) due to changes in ecological conditions as the result of the blocking of seawater intrusion, with the Diama dam (BDPA et al., 1995).

The impact of the dams on people's health has become the major problem for the whole basin since 1993/1994. A rapid increase in the prevalence of hydric diseases that were already known in the area (malaria, urinary bilharzia, diarrhoea, intestinal parasitosis) but also the appearance of intestinal bilharzia, of a more dangerous form than that which had previously existed (urinary bilharzia).

As an example, the frequency of these endemic illnesses before Manantali dam construction was as follows (Anne et al., 1994):

- malaria: 53.4%
- bilharzia: 20.2%
- diarrheic diseases: 65.9%
- ankylostomiasis: 43.4%

The installation of the dams has been accomplished without taking into account the other important aspects of planning:

- top-down planning has occurred without consultation regarding the local needs of the beneficiaries;
- the large schemes for groundnuts, cotton and irrigation have been less than successful due to the use of inappropriate technologies, the lack of markets or access to markets, and the lack of local know-how.

What is observed is that the degradation of the basin ecosystems has affected the local populations at different levels. For example, there has been a decrease in productivity in some main sectors of economic activity (agriculture, fisheries, livestock), compared with the first five years of dam filling. This has reduced the per capita income, and consequently worsened life conditions.

As an aid for analysis, prevention and mitigation of the negative impacts of the program, an Environmental Observatory was recently created. This institution is meant to be a focal point for all scientific and technical information regarding the environmental situation of the Senegal River Basin and its issues in order to provide a solid basis for the decision-making process.

Conclusions

The OMVS has demonstrated its effectiveness. It has been tested for more than thirty years now, and was recently improved by the adoption in May 2002 of the Senegal River Water Charter. This framework enables a collaborative management approach, with effective involvement of local actors/stakeholders, recognized and accepted by all the riparian states including Guinea, who has signed cooperation agreements with the OMVS prior to being reintegrated into the organization.

It also establishes the principles and terms of water sharing between the different usage sectors, based on the original concept of "water distribution" among the users and riparian states in which sharing the water resources is not a matter of withdrawals, but rather one of optimal satisfaction of usage requirements. The OMVS looks for an equitable management and distribution of water resources among multicultural ethnic groups who are gathered around the river basin and its water resources. The new dams and institutional framework have brought greater prosperity and economic revenue and replaced the situation of water scarcity and conflict among users which prevailed before the 1980's.

However, these undeniable achievements cannot hide new difficulties emerging with the dams' implementation: displacement of populations, water-borne diseases, proliferation of invasive aquatic vegetation, degradation of cultivated land and water pollution are the major issues the basin will have to cope with in the near future.

One of the main aspects that the organisation should focus on is the ecological importance of the river's floodplain which covers about 1,000,000 ha (Reizer, 1986), because its existence and development will be in conflict with the hydraulic management of the river from the Manantali dam hydro-electric power plant.

Deep reflexion is already needed in order to implement the sound management of natural resources with a view to the sustainable development of the river basin, to avoid worsening conflicts between economics and ecology.

For those who know about the geopolitics of the sub-region, one must hope that the river management will induce more communication between local populations and by doing so, consolidate peace in the region.

Acknowledgements

This work benefited from the author's previous professional involvement as an OMVS project officer. The views expressed are solely those of the author and do not represent the views of the reviewers or the institutions they represent.

References

Anne, I.; Lelek, A. and Tobias, W. (1991). African man-made lakes: critical notes on their ecology and economic contribution. *Natural Resources and Development* 33: 7–19.

Anne, I.; Lelek, A. and Tobias, W. (1994). Post impoundment changes in water quality and fish stocks in two large west african reservoirs (Manantali and Sélingué, Rep. of Mali). *Int. Revue Ges. Hydrobiol.* 79: 61–75.

BDPA-SETAGRI/ORSTOM/SECA/AFID/SERADE (1995). *Etude des problèmes d'environnement et de protection des milieux naturels dans le delta du fleuve Sénégal. Synthèse générale.*

COYNE ET BELLIER/FITCHNER/TECSULT (1996). *Synthèse des évaluations environnementales et orientations relatives à la definition du Programme d'atténuation et de suivi des impacts sur l'environnement (PASIE).*

Global Water Partnership (GWP) and International Network of Basin Organizations (INBO) (2009). *A handbook for integrated water resources management in basins.*

Obsevatoire de l'environnment. Haut Commissariat (2006). *Rapport annuel sur l'Etat de l'Environnement et des Ressources Naturelles du Bassin du Fleuve Sénégal.* Dakar.

Reizer, C. (1977). *Note technique. Etude des incidences sur l'environnement de l'aménagement du lac de barrage de Manantali sur le Bafing-Sénégal.*

Reizer, C. (1986). Impact sur l'environnement des aménagements hydrauliques en Sahel fluvial: faut-il construire les barrages sur le Sénégal? *Proceedings of Internat. Symp. UNEP-UNESCO-EMINWA-IHP*: 267–284.

Web page of the Observatoire de l'environnement. Haut Commissariat. OMVS, <www.omvs-soe.org/>.

PART V
Water and food nexus

Rainwater management in Africa: a green revolution to increase agricultural production

Nico H. van Leeuwen[*]

Rainwater management in Africa: A green revolution to increase agricultural production

Abstract. The world food crisis has deteriorated during the last few years due to substantial price increases of major food products and, more recently, because of the financial crisis. The positive trend towards the reduction of the number of undernourished people has reversed.

In sub-Sahara Africa, agriculture provides the livelihood for 70% of the population, while 93 % of the cultivated land is rain fed and 38 % of the land is located in arid or semi-arid zones. Agriculture suffers from low and erratic rainfall, resulting in low yields and frequent crop failure. Improved management of rainwater can result in substantial production increases.

Studies of the repartition of rainfall water at the soil surface show that during rainstorms only a small percentage of rainfall actually contributes to the water needs of the crop while large percentages are lost. Rainwater management technologies prevent the loss of rainwater, increase the infiltration rate and improve the water holding capacity of the soil.

Conservation Agriculture, being a package of zero-tillage, permanent soil cover and crop rotations, increases the infiltration capacity of the land and the water-holding capacity of the soils and can be applied where average annual rainfall covers the requirements of the crop. Where average annual rainfall is insufficient, rainwater-harvesting techniques are more appropriate.

The introduction and/or expansion of appropriate rainwater management practices will increase agricultural production, and additionally increase groundwater resources, protect the soil against erosion and reduce flooding and flood related damage. Further it will alleviate the negative impact of future climate change on water resources.

* Civil Engineer. Former FAO specialist in irrigation and water resources management.

Keywords. World food crisis, smallholder farmers, management of rainwater, conservation agriculture, rainwater harvesting.

Introduction

The world population continues to rise and is expected to grow from 6.9 billion today up to some 9 billion by the year 2050. Within this framework, the international community has undertaken to reduce, by the year 2015, the number of hungry people to 420 million, in other words, half of what is was in 1996 when the first World Food Summit took place. While some progress has been achieved in a number of developing countries, serious increases of undernourishment have occurred in others. In particular in sub-Saharan Africa, sharp rises of food prices in 2007–2008 and again in 2010–2011 have increased the total number of hungry people. The more recent financial crisis has aggravated the situation to the extent that the number of undernourished people exceeded the one billion mark during the course of 2009.

The increase of food prices did not originate from an actual global food shortage. At the macro level, there is still enough food for all. Changes in agricultural policies, occasional production shortages in cereal-producing countries, the increase of petrol prices, combined with a potential shift towards the production of bio-fuel and the introduction of trade policies have all contributed to the price increases. At the micro, subsistence level of the smallholder farmer, the issue is not only the availability of food but also the access to it.

In good years, subsistence farmers produce enough food to feed their families with a small surplus to be sold. In bad years they have to rely on food aid, find off-farm jobs and suffer from hunger. Food for the large cities is frequently imported from abroad. African agriculture has not changed much over the last 50 years. In order to feed the increasing population, the cultivated area has been doubled whilst the yield per unit of land has hardly changed. To reduce hunger, more food needs to be produced, firstly to provide a more stable livelihood for the rural population and secondly to feed the cities.

Several options have been proposed to increase food production but they all require ensuring a more appropriate supply of water to crops. Irrigation, the obvious solution, will not be possible in most cases for reasons of water availability and high investment costs. Consequently there are no other options than doing more with the currently available water resources, i.e. rainwater. The analysis of what actually happens to rainwater on the farmer's field indicates ways and means to retain more water in the root zone of the crop, resulting in significant increases in yields. Conservation agriculture and rainwater harvesting are examples of improved rainwater management.

Under-nourishment in sub-Saharan Africa and water resources

For many years, experts have debated the capacity of the world's agricultural systems to produce enough food for an increasing population. FAO has consistently maintained that enough food could be produced for the much larger human population predicted in 30 years time. Indeed, during the last 40 years the world population has doubled while during the same period the number of undernourished people has decreased both in absolute and in percentage terms.

At the World Food Summit in 1996 the international community undertook the commitment to, by the year 2015, reduce by half the number of undernourished people from the 1996 figure of 800 million (FAO, 1996). This commitment was reconfirmed during the Millennium Conference and at other more recent international conferences.

At the global level, the number of undernourished people decreased during the decade 1995–2005 by some 4%, raising the expectation that world food production had finally outpaced population growth and that malnutrition and hunger were in slow but steady decline. However, rising food prices in more recent years have increased again the number of hungry people by some 75 million. The most recent projections (FAO, 2009) announce a historical high of hungry people exceeding the one billion mark. The world economic crisis, combined with the recent increase in food prices, has resulted in increased unemployment and has consequently reduced access to food by the poor.

The proportion of the population suffering from under-nourishment is the highest in sub-Saharan Africa with one in three people chronically hungry. During the period 1990–2005, Sub-Saharan Africa's population grew from 500 to 700 million, placing a heavy burden on the insufficiently developed agricultural sector (FAO, 2008). In spite of the increase of the absolute number of undernourished people, the proportion of hungry people fell from 34 to 30 % during the mentioned period but increased again during recent periods as a result of high food prices.

The Food and Agriculture Organization of the United Nations has estimated that 50% more food will need to be produced by the year 2030 in order to reach the millennium goals. Figure 1 gives the three main approaches to increasing food production.

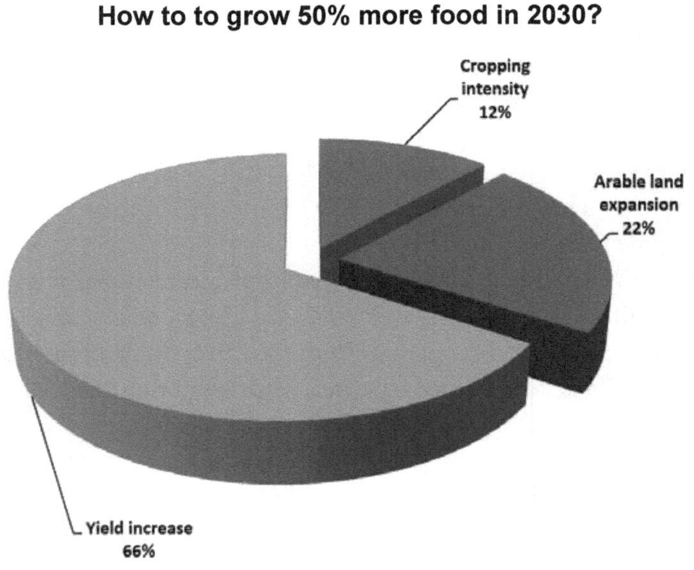

Figure 1. How to increase food production.

Yield Increase	•Adequate water supply to crops through irrigation, water harvesting/runoff farming. •Improved seed and OGM •Fertiliser •Agricultural practices
Arable land expansion	•Irrigation of dry areas •Expanded use of wetlands •Water harvesting on arid lands
Cropping intensity	•Expansion and improvement of irrigation •Reclamation and intensive use of wetlands

Table 1. Options for increased food production and the role of water.

Table 1 illustrates that any option to increase agricultural production relies strongly on the appropriate supply of water to the crops. In view of present and future water shortages, some suggestions follow:

- better use of surface and groundwater resources for irrigation: introduction of drip irrigation and other highly efficient irrigation systems, use of plastic mulch, alternate furrow irrigation, deficit irrigation, and supplementary irrigation all aiming at "more crop per drop"
- use of existing, alternative, more drought resistant crops requiring less water and the potential development of genetically modified plant material requiring less water and better adapted to endure occasional droughts
- development of crops adapted to grow under saline conditions.

When drawing attention to the importance of irrigation for food production, it is frequently quoted that 20 % of the arable, irrigated land produces 40% of the world's food crops. However, this implies at the same time that 60% of the food crops are grown under rain fed conditions on 80% of the land. Rain fed agriculture produces 69% of the world's cereal crops. Relatively small increases in the yields of rain fed agriculture will produce large increases in food production (FAO, 2003). Given the importance of rain fed cereal production, insufficient attention has been paid to the potential production growth in rain fed areas.

Dry lands constitute a significant part of the world's land resources. Following the FAO classification as indicated in table 2, 45% of the global landmass can be classified as arid. The extent of dry lands varies between regions: 38% in sub-Saharan Africa, 20% in North Africa and the Middle East, etc.

Dry-land areas are fragile because they are extremely vulnerable to land degradation, resulting from different forms of inappropriate land use. Approximately 70% of these dry lands are already degraded or subject to heavy degradation. The population living in these areas is, in most cases, in a critical situation and is using all available resources that might contribute to their survival without having the luxury of considering sustainability issues or investment in the conservation of natural assets for the future. Deforestation, over-grazing and cultivation of steep slopes are the most common sources of degradation of dry lands.

Classification	Length of growing period for annual crops in days*	% of world land area	% of land area Sub-Saharan Africa
hyper-arid	0	19	24
arid	1–59	7	6
semi-arid	60–119	20	13
dry-sub humid	120–179	18	19

* The growing period begins when the monthly precipitation exceeds half of the monthly potential evapotranspiration (FAO, 2000)

Table 2. Definition and distribution of dry land areas.

The actual net potential arable land area in sub-Saharan Africa is estimated at 10.5 million km², out of which more than 9 million km² is classified as dry land. Only 15% of the potential arable land was actually in use in 1994 (FAO, 2000). A very large potential remains unexploited but most of it is located in the arid zones where unreliable rainfall is a major risk. The livelihood of 90% of the farmers depends on rain fed agriculture (SIWI, 2000). The rainfall problems in arid and semi-arid areas are dual: the low total rainfall during the cropping season and its irregularity during that same season. As a general rule one can observe that the lower the rainfall is in a particular region, the more irregular it will be in space and time.

Finally, the rainfall variation over the crop season results in poor crop water availability, reducing rain fed yields to 25–50 % of potential yields (UNEP-SEI, 2009). The average rain fed cereal yield in developed countries in 1995 was 3.2 metric tons per hectare, virtually as high as irrigated yield in developing countries. Rain fed grain yields in developing countries are on average 1.5 metric tons per hectare. This yield gap demonstrates the high potential for the upgrade of rain fed agriculture in developing countries through improved management and use of rainwater. The following paragraphs analyse in more detail what happens with rainwater when it reaches the surface of the land and propose options to make better use of this resource.

Rainwater management, appropriate and affordable technologies

Given the importance of arid and semi-arid land for agricultural production in sub-Saharan Africa, where most of the land is rain fed cultivated and represents the only livelihood of smallholder farmers, a detailed study of the actual destination of rainwater provides indications of how to make better use of this unique supply of freshwater for agricultural use.

Analyses of water balances on farmer's fields have shown (figure 2) that when rainwater (R) reaches the soil surface, part of it infiltrates into the subsoil. The water that cannot infiltrate immediately runs off, down the slope (R off), or accumulates in small depressions at the surface of the soil from where it evaporates (Es). Of the water that infiltrates into the subsoil, part of it is used by the crop (Ec) and another part infiltrates into the soil beyond the root zone (D); it cannot be used by the crop but will contribute to the recharge of groundwater.

Research has shown (figure 3) that, with traditional dry land farming (rain-fed agriculture), 70–85% of actual rainfall is not used by the crop; this water is lost through evaporation, runoff and deep infiltration.

It is estimated that, under such conditions, 10–30% of the rainfall infiltrates into the ground beyond the root zone and contributes to the

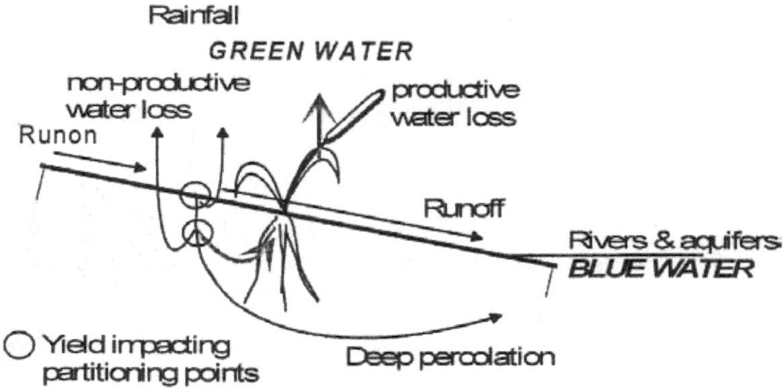

Figure 2. General overview of rainfall flows in semi-arid regions (Source: Rockström, Fox, Persson and Falkenmark, SIWI, 2000).

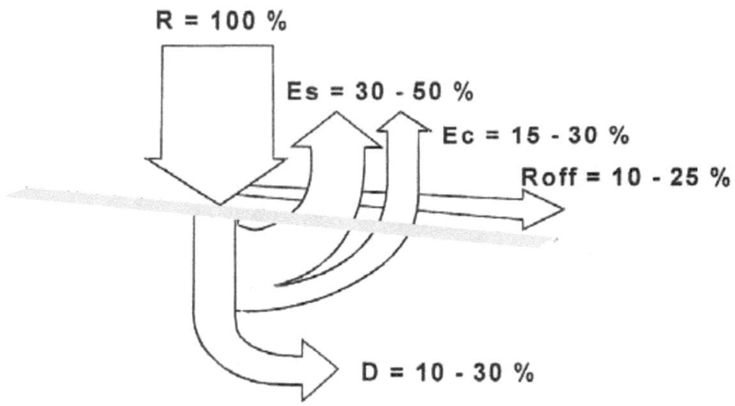

Figure 3. Repartition of rainfall flows in semi-arid regions (Source: Rockström, Fox, Persson and Falkenmark, SIWI, 2000).

recharge of groundwater. Direct evaporation from the land surface is estimated at 30–50 % while runoff amounts to 10–25 %. For mechanized agriculture, the figures might be slightly different. The challenge is to make better use of the available rainwater, and, in particular, to reduce the surface runoff, for an increase in agricultural production. It should be noted that the water not being used immediately by the crop is not necessarily entirely lost. Runoff may be captured at a lower level and used for irrigation. However, in the meantime this runoff has caused erosion upstream and sedimentation downstream and it may have contributed to flooding. Part of the runoff will also have evaporated before it can be collected again. Deep-percolation of rainwater is recharging the groundwater resources and will contribute either to regulating river discharge or can be pumped-up for domestic and industrial use or for irrigation.

In many climatic zones, and in semi-arid areas in particular, it is expected that future rainfall, due to climate change, will be less frequent but of a higher intensity. This will result in higher runoff with the corresponding negative consequences.

Conservation agriculture (CA) and rainwater harvesting (RWH) reduce the volume of surface runoff and they increase the quantity of water available for crops. Compared to irrigation, both technologies have significant advantages. To begin with, it should be noted that, because of the lack of accessible water resources, most arid and semi-arid land cannot be irrigated. Further, as has been experienced in many places, irrigation systems, and in particular large-scale irrigation projects, are very costly to install and to maintain. They also require all participating farmers to respect specific rules (amount of water, time schedule, etc) for the appropriate operation of the system. As an alternative to irrigation, the above mentioned rainwater management technologies are much more accessible as activities can be undertaken individually. Each farmer is free to apply them on his land independently of the size of the landholding or the possible participation of his neighbours. Farmers can select the most appropriate package of technologies, adapted to their local circumstances, which is also of particular importance for women farmers. There is a lot of flexibility as farmers can expand (or reduce) the land area covered and they can also change easily to other technologies.

Conservation agriculture (CA) and rainwater harvesting (RWH) are affordable technologies because:

– initial investment costs are low. Small-scale implementation of these technologies requires mainly additional labour that can be supplied by the farmer and his family during low-activity periods. Farm machinery, possibly required for larger scale implementation of conservation agriculture, is generally already available.

- operation and maintenance costs are low. While there are hardly any operating costs (water will be supplied at no cost by the rain if and when it comes) water-harvesting structures need occasional maintenance in order to ensure they remain effective.
- ease of installation and operation. While the construction of rainwater management structures is relatively easy, there may be a need to provide some initial assistance to smallholder farmers in regions where traditional knowledge is lacking. Once a number of demonstration plots are operational, farmers will easily learn from others how to apply these technologies in their own fields.
- cost-effectiveness. All trials and pilot projects have yielded a significant and immediate increase in agricultural production, even for crops grown the traditional way (local seed, no fertilizer, etc.). Yield increases of at least 100% are mentioned in most cases. A most interesting feature is that rainwater management reduces the risk of crop failure and encourages farmers to invest in other inputs such as fertilizer and improved seed that further increase yields.

Conservation agriculture

Conservation agriculture (CA) is a concept for resource-saving agricultural crop production that strives to achieve acceptable profits together with high and sustained production levels while concurrently conserving the environment (FAO, 2006). It is based on enhancing natural biological processes above and below ground. Interventions such as mechanical soil tillage are reduced to an absolute minimum, and the use of external inputs such as agrochemicals and nutrients of mineral or organic origin are applied at an optimum level and in a manner and quantity that does not interfere with, or disrupt, the biological processes in the soil. Conservation agriculture is characterized by three interlinking principles, namely:

- Continuous minimum mechanical soil disturbance (zero tillage)

- Permanent organic soil cover.
- Diversified crop rotations in the case of annual crops or plant associations in the case of perennial crops.

Conservation agriculture, understood in this way, provides a number of advantages on the global, regional, local and farm levels. One of these relates particularly to water resources. Soils under conservation agriculture have very high water infiltration capacities, thus significantly reducing surface runoff and thus soil erosion. This improves the quality of surface water, reducing pollution from soil erosion, and enhances groundwater resources. In many areas it has been observed after some years of conservation farming that natural springs, which had dried up many years previously, have started to flow again. The potential effect of a massive adoption of conservation farming on global water balance is not yet fully recognized at all levels.

Conservation tillage started in the early sixties and seventies in the USA and Brazil. After a slow start, it is now spreading rapidly especially in North and South America and in Australia. In 2004–2005 some 95 million ha were under no-tillage: 40% in North America, 47% in Latin America and 9 % in Australia. It has also been introduced on small areas in Spain and in France (FAO, 2006). Conservation agriculture uses adapted machinery and it is well suitable for mechanized large-scale farming. It is, however, only effective in those regions where rainfall is sufficient to grow one or more crops during the season, unless irrigation can be applied.

The adoption of conservation agriculture practices in areas with appropriate rainfall often leads to increased yields (double or even triple). Among the other major advantages is the reduction in the amount and costs of labour and energy required for land preparation that often constitutes an important bottleneck. It has also been demonstrated that the improved soil conditions make the land and the agricultural system more resilient to extreme events (FAO, 2001).

Rainwater harvesting

In areas where rainfall is erratic and where the seasonal rainfall is insufficient to grow crops, rainwater harvesting can be applied. Water harvesting in its broadest sense can be defined as a "collection of runoff for its productive use". More precisely, water harvesting is the process of collecting and concentrating rainfall as runoff from larger non-cultivated areas to be used in a smaller cultivated area. The collected water is stored in the soil profile for immediate uptake by the crop or stored in a reservoir for future productive use (domestic supply, livestock, aquaculture and irrigation). Water harvesting systems are characterized by the presence of a run-off area and a run-on area.

The run-off area can be any impervious compacted smooth area. Inside the cropped area, parts of the land can be kept clean from vegetation and treated to increase the run-off. Also, the water running-off from catchments outside the cropped area can be deviated towards this area, in other words, spate irrigation through floodwater diversion. The run-on zone is the cultivated area and in most cases water is stored in the root zone. Excess water will join the shallow groundwater aquifer and can be pumped

Figure 4. Rainwater harvesting Keita project Niger. (from FAO, 1995).

up for irrigation. Rainwater harvesting is not an innovation. For many centuries, a large number of different techniques have been successfully used worldwide to obtain optimal use of scarce rainwater resources. The selection of one particular technique depends on the rainfall, the slope and nature of the soil and the specific crop requirements (FAO, 2003.a). Different techniques are used for cereal crops, fodder, trees, etc. Many of the techniques are labour intensive and adapted to small farms while others, such as the Vallerani technique, are mechanized and are used on large areas (Antinori, 2001).

Where small areas are treated to retain and harvest rainwater, the impact on agricultural production is immediate. When larger areas are treated (figure 4) the additional impact on global water resources becomes evident. An example of such a large area is the Keita district in Niger where, with the assistance of the Italian Government through FAO, an area of 320,000 ha was treated for rainwater harvesting (FAO, 1995). This region has a typical Sahelian climate with average annual rain of 300–400 mm/year falling sporadically during the period June – September. Initially the area was subject to severe erosion and desertification. As a result of the treatment of large areas for the collection and retention of rainwater, the area has now become green, agricultural production has significantly increased and, with particular regard to water resources, shallow wells and boreholes are providing water all year round for domestic purposes and for irrigation. Similar results have been obtained in many other arid and semi-arid regions.

Conclusions

Extensive areas of sub-Saharan Africa are situated in arid and semi-arid zones where rain fed agriculture is the main livelihood of smallholder farmers. Because of low and erratic rainfall, yields are low and depend to a large extent on the distribution of rainfall during the crop season. There is a large gap between the actual farmer's yields and the achievable yields for the major rain fed crops, mainly due to the lack of humidity in the root zone at critical periods of the crop season.

Studies of the movement of rainwater at the surface of the soil during rainstorms have shown that large amounts of rainwater are lost for agricultural production or for groundwater recharge because of surface runoff and direct evaporation. Higher rainfall intensities, expected as a result of future climate change, will exacerbate such water losses. Efforts to augment food production in the arid and semi-arid zones should focus on the increase of water supply to the crops through improved management of rainwater.

In areas with rather favorable rainfall conditions, conservation agriculture techniques have demonstrated an immediate and significant impact on agricultural yields, increasing at the same time the global availability of water. In arid and semi-arid areas, ancient rainwater harvesting techniques, re-discovered and adapted to modern conditions (mechanization), have demonstrated similar impacts.

The introduction and/or expansion of agricultural practices aiming at improved rainwater management can dramatically increase agricultural production, reduce irrigation water requirements, increase groundwater resources, protect the soil against erosion, and will indirectly reduce downstream flooding and flood-related damage. The potential of rainwater management is such that it can undoubtedly be described as a green revolution, however with significantly less damage to the environment compared to the first such revolution.

References

Antinori, P., Vallerani, A, and Vallerani, V. (2001). *The Vallerani System, a technology specially conceived for fighting against desertification, hunger and poverty.* Private communication, printed with the contribution of the Italian Cooperation, 2002, Rome.

FAO (1995). The Keita Integrated Rural Development Project. FAO, Rome.

FAO (1996). *The Rome Declaration on World Food Security and World Food Summit Plan of Action.* Rome: FAO.

FAO (2000). *Land Resources Potential and Constraints at Regional and Country Level.* World Soil Resources Report 90. Rome: FAO.

FAO (2001). *Conservation Agriculture. Case studies in Latin America and Africa. Land and Plant Nutrition Management Service.* FAO Soils Bulletin 78.

FAO (2003). *Unlocking the Water Potential of Agriculture.* Rome: FAO.

FAO (2003a). *Training Course on Water Harvesting.* Land and Water Digital Media Series, CD Rom N° 26. Rome: FAO.

FAO (2006). Fact Sheet on Conservation Agriculture, <www.fao.org/ag/ca/1a.html> (accessed on 29 Jun 2009).

FAO (2008). *The State of Food Insecurity in the World.* Rome: FAO.

FAO (2009). 1.02 Billion people hungry, <www.fao.org/news/story/en/item/20568> (accessed on 29 Jun 2009).

SIWI (2000). Rockström, J. , Fox, P. , Persson, G. and Falkenmark, M. Water. *Harvesting for Upgrading of Rainfed Agriculture. Problem Analysis and Research Needs.* Stockholm, Sweden.

UNEP/SEI (2009). *Rainwater Harvesting: a Lifeline for Human Well-Being.* Report prepared for UNEP by Stockholm Environment Institute Editor: Jennie Barron, Stockholm Environment Institute, York, UK/Stockholm Resilience Centre, Stockholm, Sweden.

Notes on contributors

DANIEL ALVARENGA. Former analyst, Centre for Chinese Studies at Stellenbosch University (South Africa).

IBRAHIMA ANNE. Born in Mauritania in 1962, studied in his country, Sénégal, Ivory Coast, Russia, Ukraine (Master thesis in hydrobiology) and Portugal (Doctoral thesis in environmental microbiology); training courses during his doctoral studies in Germany (Technical University of Munich), Denmark (Agricultural University of Copenhagen) and France (Université de Clermont Ferrand). Member of the Societas Internationalis Limnologiae, and of the Portuguese Association of Biologists. Currently works at the technical division of Degrémont SA, where in his duties as technical assistance manager, has visited Algeria, Egypt, Mauritania and Russia.

ADRIANO A. BORDALO is an Associate Pofessor at the Intitute of Biomedical Sciences, Porto University (ICBAS-UP), and the head of of the Laboratory of Hydrobibology. He published over 65 peer-reviewed scientific papers and over 200 communications on different environmental subjects, some of them dealing with Africa.

HELEN BROWN. Department of Geography, Sheffield University. Her interests include social learning within participatory, catchment-based water management processes in South Africa and Namibia.

MANUEL C. BRANCO. Associate professor of economics at the University of Évora, Portugal. He graduated in Economics and in Geography and received his Phd in Economics. His main research interests concern the political economy of development and underdevelopment and the interaction of economic and non-economic factors in the development process, more specifically with respect to ethics, democracy, culture and inequality. His latest publications include the book Economics versus Human and the articles The Political Economy of the Human Right to Water and Economics Against Democracy.

SIGNE MARIE COLD-RAVNKILDE is a researcher at the Danish Institute of International Studies (DIIS). Her worked has focused on natural resource conflicts and institutions in West Africa including Struggling for water and pastures in Niger: Natural resource conflict and cooperation in the pastoral areas of the Diffa region, 2009. Her thesis 'A Gift from God: Conflicts over Water and Authority in Mali' (forthcoming) addresses local government institutions establishing authority through control over strategic water resources in a context of democratic decentralization.

PEDRO DAMIÃO DE SOUSA HENRIQUES is an Engineer in Agronomy, with a Master and a PhD in Agricultural Economics. He is an Associate Professor at the Department of Economics, University of Évora, Portugal. His main research interests are in the field of agricultural and resource economics and policies such as: agricultural efficiency; water policy and human rights; land, gender and development. He has participated in several research projects and in co-operation programmes for development, especially with Angola and Timor-Leste.

NICO VAN LEEUWEN. After having obtained a M.Sc. in civil engineering with specialization in hydraulics at the Delft Technical University, Nico van Leeuwen was employed by the Food and Agriculture Organization (FAO) of United Nations for 32 years until retiring in 2004. During an initial period of 14 years he worked in technical cooperation projects in Niger, Lebanon, Cape Verde Islands and Burundi and acquired a thorough field experience of all water related aspects of agricultural development in Africa. After having managed complex development projects in the same continent for a number of years, he was appointed senior officer at the FAO Water Resources, Development and Management Service where he focused on the promotion of small-scale, low-cost water resources development technologies for smallholder farmers such as: small-scale irrigation, rainwater harvesting, wetland development and management, shallow groundwater development and management, etc. in particular for dry-land areas in Sub-Saharan Africa.

STEFAN LIEHR is a senior researcher at the Institute for Social-Ecological Research (ISOE) in Frankfurt am Main/Germany. He is leading the research unit "water resources and land use" and is involved in different projects and activities in Namibia focusing on integrated water resources management. His disciplinary background is Physics, Chemistry and Geophysics.

PAULA DUARTE LOPES is a researcher on Peace Studies at the Center for Social Studies and an International Relations professor at the School of Economics of the University of Coimbra (Portugal). She lectures at the undergraduate and graduate levels in International Relations, including the PhD Programme in International Politics and Conflict Resolution. She received her PhD from The

Johns Hopkins University (United States of America), and her Masters from the London School of Economics and Political Science (Great Britain). Her research interests include peace studies, environmental governance, international water politics and development aid. She has several publications and papers on the dynamics of water and environmental international governance, which you can check at <www.ces.uc.pt/investigadores/cv/paula_duarte_lopes.php>.

Agathe Maupin has completed a Philosophical Degree in Tropical Geography at the University of Bordeaux 3 in 2010. Some of the results of the author's work in Southern African transboundary River Basins, conducted from 2006 till 2010 on the Southern African Hydropolitical Space, are presented here. Several publications have already been taken from her PhD work (Sécurité Globale, BAGF, Les Cahiers d'Outre Mer). The author, also agrégée of Geography, has taught as a temporary Research and Teaching Assistant at the University of Bordeaux 3, then of Paris IV-Sorbonne. She is now working at the South African Institute of International Affairs (SAIIA) as an expert on the Energy Security impacts on International Relations in a global context of Climate Change.

Godwin Uyi Ojo. Sessional Lecturer, Department of Geography, School of Social Science and Public Policy, King's College London.

Mathias Polak is a Ph.D. candidate at the University of Frankfurt, Germany. He is working on water governance and institutional change in central Nortern Namibia. His research is part of the project "CuveWaters: Sustainable Water-Management in Namibia".

Joana Savva-Bordalo is a Medical Doctor (Oncology) with clinical research experience in Guinea-Bissau. Besides several peer-reviwed scientific papers, she is a co-author of over a dozen on water quality in Africa.

CEAUP Studies on Africa

Vol. 1 Alexander Keese (ed.)
Ethnicity and the long-term perspective. The African experience.
ISBN 978-3-0343-0337-8. 2010.

Vol. 2 Ibrahima Anne (ed.)
Water Governance – Challenges in Africa.
Hydro-optimism or hydro-pessimism?
ISBN 978-3-0343-1133-5 pb., ISBN 978-3-0351-0518-6 eBook. 2013.

FSC
www.fsc.org
MIX
Papier | Fördert gute Waldnutzung
FSC® C083411

Kolibri 360 GmbH
Ferdinand-Jühlke-Straße 7
99095 Erfurt, Deutschland
produktsicherheit@kolibri360.de